ULTRASONIC SECTIONAL ANATOMY

To Professor Ian Donald—who first introduced me to Ultrasound and whose infectious enthusiasm inspired my own interest in the subject.

PATRICIA MORLEY

'Enthusiasm is the genius of sincerity, and truth accomplishes no victories without it'.
    LYTTON—*The last days of Pompeii*

ULTRASONIC SECTIONAL ANATOMY

Edited by

Patricia Morley
MB BS DMRD FRCR
Consultant Radiologist, Ultrasonic Unit, Department of
Diagnostic Radiology, Western Infirmary, Glasgow

Gabriel Donald
DA FMAA
Senior Lecturer, Department of Medical Illustration,
Western Infirmary, Glasgow

Roger Sanders
MD
Associate Professor, Department of Radiology and
Radiological Sciences, The Johns Hopkins Hospital,
Baltimore

CHURCHILL LIVINGSTONE
EDINBURGH LONDON MELBOURNE AND NEW YORK 1983

CHURCHILL LIVINGSTONE
Medical Division of Longman Group Limited

Distributed in the United States of America by Churchill Livingstone Inc., 1560 Broadway, New York, N.Y. 10036, and by associated companies, branches and representatives throughout the world.

© Longman Group Limited 1983

All rights reserved. No part of this publication may be reproduced, stored in a retrieval system, or transmitted in any form or by any means, electronic, mechanical, photocopying, recording or otherwise, without the prior permission of the publishers (Churchill Livingstone, Robert Stevenson House, 1–3 Baxter's Place, Leith Walk, Edinburgh, EH1 3AF).

First published 1983

ISBN 0 443 01690 9

British Library Cataloguing in Publication Data
Morley, Patricia
 Ultrasonic sectional anatomy
 1. Ultrasonics in medicine
 I. Title II. Donald, Gabriel III. Sanders, Roger
 610'.28   R587.U48

Library of Congress Cataloging in Publication Data
Morley, Patricia.
 Ultrasonic sectional anatomy.
 Includes index.
 1. Diagnosis, Ultrasonic. 2. Anatomy,
Human. I. Donald, Gabriel. II. Sanders, Roger
C. III. Title.
 RC78.7.U4M67 1983   616.07'543   81-71720
                                     AACR2

Printed and bound in Great Britain by
William Clowes (Beccles) Limited, Beccles and London

PREFACE

The aim of this book is to provide a systematic series of sections basic to ultrasonic anatomy. Other texts concentrate on pathological features; however it is obvious that a thorough knowledge of the normal has to be acquired before an attempt is made to interpret the abnormal image.

The majority of the ultrasonic sections have been recorded during routine examination of patients referred for ultrasonic diagnosis. Tubular structures have not been selectively recorded in the black echo format nor textural detail as white echoes. The scans are therefore of the quality seen daily in a busy diagnostic department and include normal variants that have been encountered by the authors. The sections are recorded in the recognised anatomic planes, and with few exceptions oblique scans have been deliberately avoided.

With the continuous development in ultrasonic technology fresh information continues to evolve and no textbook on ultrasound can be fully comprehensive by its date of publication. We do, however, appear to be reaching a plateau, and though some of the images are dated, I would not anticipate any major advance in the near future. However, 'He that would know what shall be must consider what has been' (*Proverbs. Gnomologia*—Dr Thomas Fuller, 1732) and, reflecting on the advances that have taken place in the last decade, anything becomes possible.

The scope of the book should be adequate for the requirement of candidates for the Fellowship of the Royal College of Radiologists, the Diploma in Medical Ultrasound of the College of Radiographers, the American Society of Ultrasound Specialists, and the American Boards of Radiology. It should also be of value to medical students who now have cross-sectional anatomy included in their curriculum. I also hope that it will be used by other medical specialists who wish to acquire the ability to understand the images of the ultrasonic investigations that they are now requesting in ever-increasing volume.

My debt to the other contributors is great—they have vastly increased the range of the book. Particularly I must thank Mr Gabriel Donald and Mrs Priscilla Miles who have prepared the line drawings—their patience with my vacillation has been remarkable.

Glasgow, 1982   P.M.

CONTRIBUTORS

A. Hunter Adam MB ChB MRCOG
Senior Registrar, Department of Obstetrics and Gynaecology, University of Aberdeen, Aberdeen Maternity Hospital, Scotland

Ellis Barnett FRCR FRCP
Consultant Radiologist, Department of Diagnostic Radiology, Gartnavel General Hospital and Western Infirmary, Glasgow, Scotland

Gabriel Donald DA FMAA
Senior Lecturer, Department of Medical Illustration, Western Infirmary, Glasgow, Scotland

W. J. Garrett MD DPhil FRCS(Ed) FRACS FRCOG
Director, Department of Diagnostic Ultrasound, Royal Hospital for Women, Sydney 2021, Australia

J. Hackelöer MD
Professor Dr Gynaecologie, Univ-Frauenklinik, Pilgrimstein 3, D-3550 Marburg, West Germany

J. Jellins BSc BE
Technical Project Director, Ultrasonics Institute, Australian Department of Health, Sydney 2000, Australia

Malcolm LeMay MB ChB FRCS
Senior Lecturer in Ophthalmology, University of Glasgow Tennent Institute of Ophthalmology; Honorary Consultant Ophthalmologist, Western Infirmary, Glasgow, Scotland

Edward Lipsit MD
Assistant Clinical Professor of Radiology, George Washington Hospital, Washington DC, USA

Dennis McQuown MD
Medical Director, Department of Diagnostic Ultrasound, Memorial Hospital Medical Centre, Long Beach, California, USA

Patricia Morley MB BS DMRD FRCR
Consultant Radiologist, Ultrasonic Unit, Department of Diagnostic Radiology, Western Infirmary, Glasgow, Scotland

Marie Restori BSc MSc
Senior Physicist, Ultrasonic Department, Moorfields Eye Hospital, London

Hugh P. Robinson MB ChB MD MRCOG FAust COG
First Assistant, Department of Obstetrics and Gynaecology, University of Melbourne, Royal Women's Hospital, Melbourne, Australia

J. Christine Rodger MD FRCP(Glasg)
Consultant Physician, Monklands District General Hospital, Airdrie, Lanarkshire, Scotland

Roger C. Sanders MD
Associate Professor, Russell Morgan Department of Radiology and Radiological Sciences, The Johns Hopkins Medical Institution, Baltimore, Maryland, USA

Eugene Strandness MD
Professor of Surgery, Department of Surgery, University of Washington, Seattle, Washington, USA

James L. Weiss MD
Assistant Professor of Medicine, The Johns Hopkins Medical Institution, Baltimore, Maryland, USA

Mark Ziervogel BSc MB ChB FRCR
Consultant Radiologist, Department of Diagnostic Radiology, Royal Hospital for Sick Children, Yorkhill, Glasgow, Scotland

Acknowledgements

The following figures have been reproduced in this text with the kind permission of the respective authors, editors and publishers.

Figures 3.1 and 3.5B: from Morley P, Barnett E, Lerski R A, Young R E In: Thijssen J M (ed) Ultrasonic tissue characterisation. Stafleu's Scientific Publishing Co., Leiden

Figures 6.5, 6.20, 7.26, 7.30, 7.62, 7.71, 7.72, 7.79, 9.11, 9.29, 11.9, 11.16, 11.40 and 11.47: from Barnett E, Morley P 1979 In: Margulis A R, Burhenne H J (eds) Alimentary tract radiology. C. V. Mosby Co., St Louis

Figures 6.9 and 10.9: from Morley P, Barnett E In: de Vlieger et al Handbook of clinical ultrasound. Wiley Medical, New York

Figures 9.1, 9.2, 9.5, 9.18, 10.49A and 12.46: from Morley P 1979 In: Rosenfield A T (ed) Genitourinary ultrasonography. Churchill Livingstone, New York

Figures 9.48 and 9.49: from Hackelöer B J, Robinson H P 1978 Ultrasound examination of the growing ovarian follicle and the corpus luteum during the normal menstrual phase. Geburtshilfe und Frauenheilkunde 38:163—166

Figure 10.33: from Miskin M Ultrasound in paediatrics. Grune and Stratton, New York

Figure 11.39: from Meyers M A Dynamic radiology of the abdomen.

CONTENTS

Introduction   xiii

1. The brain and cerebral ventricles   1
    Cross-sectional echography of the brain
    William J. Garrett
    The cerebral ventricles

2. The eye and orbit   *Malcolm LeMay*   12

3. The thyroid and adjacent soft tissues of the neck   23

4. The breast   *William J. Garrett*   27

5. The heart   *James L. Weiss and J. Christine Rodger*   35

6. The abdominal muscles and skeletal boundaries   44
    The anterior abdominal wall
    The diaphragm
    The posterior abdominal wall—supine sections
    The posterior abdominal wall—prone sections
    Skeletal boundaries and muscles of the pelvis
        Skeletal boundaries
        The pelvic muscles

7. Upper abdominal viscera   63
    The liver, hepatic veins and intrahepatic portal system
    The gall bladder and bile ducts
    The pancreas
    The spleen
    The adrenal glands

8. The kidneys   99
    The normotopic and ectopic kidney
    Renal anomalies
    The transplanted kidney

9. The viscera of the lower abdomen and pelvis   117
    The urinary bladder, ureter, seminal vesicles and prostate
    The scrotum and penis

    Examination of the penis and scrotum using the UI Octoson
    William J. Garrett and J. Jellins

10. The gastro-intestinal tract and peritoneal cavity   144

    The gastro-intestinal tract
    The peritoneal recesses

11. The major vessels   172

    The major visceral arteries and veins
    The abdominal aorta and major branches
    The inferior vena cava and major tributaries
    The extrahepatic portal system
    Vessels in the lower limb and neck

12. Obstetric ultrasound *H. P. Robinson*   196

    Basic embryology
    The form of the embryo at different stages of growth related to the ultrasonic image
    The first trimester
    Twin gestation
    Fetal anatomy
    The placenta

References   227

Index   229

INTRODUCTION

Since pulsed echo techniques were first developed in the early 1950s, the principal mode of use in the abdomen has been through a 'static' scanner with an articulated arm. The image is produced over a period of 25–30 seconds as the transducer is manipulated manually around the abdomen. When only bistable imaging was available, a compounding overriding technique was used. Two-dimensional views were composed from a number of different angles so that interfaces could add to form a complete outline of a structure. With the advent of grey scale ultrasound and the ability to see not only specular reflections from large interfaces at right angles to the ultrasonic beam but also back-scattered echoes that emanate from small structures and propagate in every direction, the use of a 'single pass' technique became of considerable importance. In certain areas such as the liver when there are suspected metastases, a 'single pass' technique is now the only satisfactory acceptable method; compounding overriding techniques obliterate small structures.

When B-scan ultrasonic images (otherwise known as static images) first became available, low-frequency transducers had to be used because of the relatively poor power output with most instruments. Sufficient penetration into the body could only be obtained with low frequencies; 1.5 and 2.25 MHz transducers were used even in thin people. More modern equipment permits the use of significantly higher frequency transducers and it is now common to perform examinations of all the organs in the abdomen in thin individuals with a 3.5 or 5 MHz transducer. Expected resolution is now 1–2 mm close to the skin and about 0.5 cm at a depth of 10–15 cm. All along, a technical problem has been the selection of the correct amount of ultrasonic power for the organ being imaged with the optimal setting of the time gain compensation (TGC)—otherwise known as swept gain. It is important to set the TGC at a setting which gives adequate information about far structures and yet does not swamp the near structures with too many echoes. One might think that standard settings could be devised for structures such as the liver, kidneys, etc., but this has not proven possible in practice because individual differences in size and amount of fat and muscle exist. A tailored TGC has to be established, even with grey scale, for every patient.

The last few years have seen the development of numerous different types of 'real-time' equipment. The first two widely used real-time instruments were not really appropriate for the abdomen outside the field of obstetrics because of their shape or expense. The 'sequenced linear array' is a system that gives a smallish square image and uses a long bar imaging device containing numerous small transducers firing in groups of four. Grouping the small transducers in fours decreases the beam spread.

Although this instrument can be satisfactory for some patients outside obstetrics, particularly in the examination of the aorta, its shape makes examinations of many organs in the upper abdomen inconvenient. It cannot be used in every direction between ribs; however, unorthodox views often achieve satisfactory results. Phased array systems with a small transducer head developed for the heart were never widely used in the abdomen because of their expense and limited field of view. A number of mechanical sector ultrasonic instruments have recently been introduced which use either an oscillating transducer or rotating wheel equipped with three or four transducers as the source of the ultrasound. The resolution achieved with these systems is fully comparable to that developed with static scanners although a permanent record with satisfactory photography still remains a problem. However, these instruments have a small field of view and it is therefore difficult to interpret a study at a later date if the examination is of a relatively large field, e.g. a large mass or a survey for nodes. Although opinion has not yet firmed up, it seems likely that such real-time scanners will be the first method of examination of localised problems such as gall bladders, kidneys for obstruction, or ascites, but that full-scale examinations of the whole abdomen will be performed with a static scanner.

In recent years manufacturers have produced ultrasonic instruments designed for specific areas. Because the eye is so superficial and filled with fluid, frequencies of between 8 and 30 MHz are used to examine it. These frequencies give exquisite resolution but are so high that they cannot be used in other parts of the body. A separate breed of ophthalmological ultrasonic instruments has evolved, either based on contact scanning through a water bath or on a real-time mechanical sector scanner with an 8–10 MHz frequency. The high frequencies used in these systems means that the electronics are currently not technically interchangeable with systems used elsewhere.

Another area in which morphological problems have led to a specific ultrasonic instrumentation is in the prostate and the base of the bladder. Transducers mounted on the end of a rod which rotates through a 360 degree axis (sometimes known as radial scanners) are surrounded by a balloon-covered water bath or placed on the end of a cystoscope. Such a system is inserted either into the rectum or into the bladder. These systems were developed because bone in the pubic symphysis prevents access to the prostate from an abdominal approach. The presence of a transducer within the bladder or rectum so close to the organ in question allows unimpeded visualisation and the use of high frequencies. The relatively limited area that can be examined with these systems has prevented their widespread use.

Specific instruments known as small parts scanners have also been developed to examine the first 4 or 5 cm below the skin. These instruments use frequencies in the 5–10 MHz range and are designed to look at superficial organs or vessels in which very high resolution is critical such as the thyroid, testicle or carotid artery. These instruments have, incidentally, been found to be of value in the examination of infants with kidney, liver or spinal problems. Vessels are also being examined with instruments which use Doppler. Doppler can be imaged in a two-dimensional form if it is pulsed and this gives information about the flow through the vessel. A combination of a small parts scanner and Doppler system is proving of value in the evaluation of carotid artery narrowing.

The breast is another area where systems devoted solely to sonographic breast analysis have been developed. These systems are expensive but are semi-automated. Several sweeps are automatically performed but alterations of the time-gain compensation, power output and the number of transducers used are made by the operator. Such systems use up to eight transducers simultaneously. Whether breast detail is better obtained with these systems rather than with a real-time or static scanner is uncertain but the technique allows the examination to be performed by a less skilled operator.

As with any imaging technique, anatomical appearances vary between different individuals and are not merely dependent on sex or size but also on the intrinsic wide variation in normal anatomical make-up. In this book we attempt to cover the spectrum of normal anatomy, the major normal variants, and the variants caused by previous surgery. An additional feature that we illustrate is the question of variants which are caused by the ultrasonic technique itself. Unquestionably ultrasound is more technique-dependent than other imaging procedures. Technical problems can be due to scanning technique, the fashion in which the sound beam strikes the patient and to partial system failure. Each of these variant situations must be recognised if an adequate interpretation of the ultrasonic examination is to be made.

Baltimore, 1982   R.C.S.

EDITORIAL NOTE

The majority of sections are displayed in the standard transverse, longitudinal and coronal imaging planes. Transverse planes are viewed as from the subject's feet. With longitudinal or sagittal planes the subject's head is on the left and feet on the right of the section. With coronal planes the subject's head is on the left with the side up at the top of the image; the side down is indicated in the legend. This is the standard adopted by the American Institute for Ultrasound in Medicine as an interim in 1976 and confirmed in 1977 (A.I.U.M. 1976). There are a few exceptions which are indicated in the text.

A line of depth markers representing 1-centimetre intervals in the subject are recorded on some of the scans.

1 THE BRAIN AND CEREBRAL VENTRICLES

CROSS-SECTIONAL ECHOGRAPHY OF THE BRAIN
William J. Garrett

B-mode examination of the brain is not suited to adults as the thickness of the skull prevents the acquisition of satisfactory echograms. Up to the age of two years, the skull does not interfere significantly with the transmission and reception of ultrasound so good quality images of both the brain tissue and the ventricular system can be obtained. From 2 to 12 years of age, the quality of results is variable. In younger members of this group the landmarks of the brain can usually be identified. As the skull thickens, brain tissue detail is lost but the ventricular outlines remain and measurement of the ventricles is possible. There are however some children who, from the age of 3 years, have thick skulls and with whom satisfactory echograms cannot be obtained.

When the examination is made with a contact echoscope (Kossoff et al 1974, Garrett et al 1975) it is necessary to examine the head from both sides as the anatomy in the near field is not adequately displayed. Transverse sections parallel to Reid's base line are limited by the ear so the inferior horns of the lateral ventricles and posterior fossa are not displayed. With a water-delay echoscope such as the UI Octoson which was used for the present work (Kossoff et al 1975, Garrett & Kossoff 1977), the whole cross-section is seen in each echogram and the examination is made from one side only. The ear is covered with contact medium and, as the transducers do not touch the patient, true transverse sections can be carried down to the level of the floor of the middle cranial fossa. With a water-delay system it is possible to tilt the plane of section 15° to 20° to Reid's base line. This brings the cerebellum more clearly into view.

In young babies a symmetrical low-level echo area is not infrequently seen peripherally extending from the temporal pole to the occipital pole (Fig. 1.1 label 33). The exact nature of this area is difficult to determine but it is certainly a normal appearance. On the near side to the transducers, the almost complete lack of echoes could be partly due to the overlying bone but this is not the whole reason as a similar area is defined on the far side where bony interference cannot be invoked. In more than 1000 B-mode examinations, this low-level echo area has only been seen in neonates and in the absence of hydrocephalus. At first sight the appearance suggests a cistern containing cerebrospinal fluid, but from its position only part of the area can be so explained. Large myelinated nerve tracts such as the optic nerve, vagus and brain stem return very few echoes. It is possible that this low-level echo area near the cortex is similarly related to the state of myelination and hydration of the nerve cells and to the cellular content and capillary vascularity of the tissue. The lateral cerebral sulcus may also contribute to the appearance.

2 ULTRASONIC SECTIONAL ANATOMY

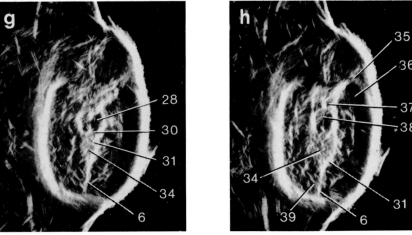
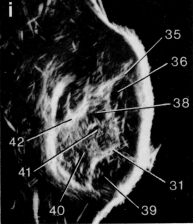

Fig. 1.1 Transverse sections of the brain.
Letters refer to plane of section shown on diagram on opposite page.

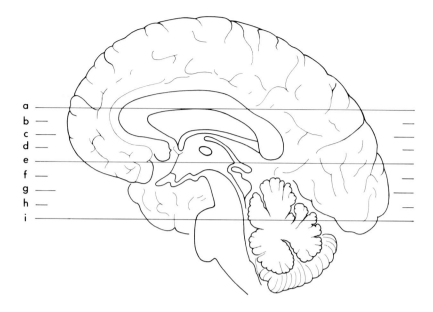

1. Falx cerebri anterior to corpus callosum
2. Body of corpus callosum
3. Body of lateral ventricle
4. Lateral wall of body of lateral ventricle
5. Atrium of lateral ventricle
6. Falx cerebri posterior to corpus callosum
7. Choroid plexus
8. Anterior horn of lateral ventricle
9. Head of caudate nucleus
10. Septum pellucidum
11. Splenium of corpus callosum
12. Forceps major of corpus callosum
13. Sulcus of insula
14. Genu of corpus callosum
15. Cavum of septum pellucidum
16. Body of fornix
17. Thalamus
18. Fronto-parietal suture
19. Foramen of Monro
20. Third ventricle
21. External capsule
22. Lateral aspect of lentiform nucleus
23. Posterior part of hippocampus
24. Choroid fissure
25. Lateral cerebral fissure
26. Olfactory sulcus
27. Straight gyrus
28. Cerebral peduncle
29. Cerebral aqueduct
30. Colliculus
31. Tentorium cerebelli
32. Posterior horn of lateral ventricle
33. Low-level echo area (see text)
34. Vermis of cerebellum
35. Lesser wing of sphenoid bone
36. Temporal lobe
37. Body of sphenoid bone
38. Pons
39. Occipital lobe
40. Cerebellum
41. Fourth ventricle
42. Petrous temporal bone
43. Cingulate sulcus
44. Stylus of septum pellucidum
45. Great cerebral vein
46. Inferior horn of lateral ventricle
47. Optic recess of third ventricle
48. Crus cerebri
49. Falx above the corpus callosum
50. Column of fornix
51. Stem of lateral sulcus★
52. Sulcus of corpus callosum

★ The term 'lateral sulcus' is synonymous with Sylvian fissure.

4 ULTRASONIC SECTIONAL ANATOMY

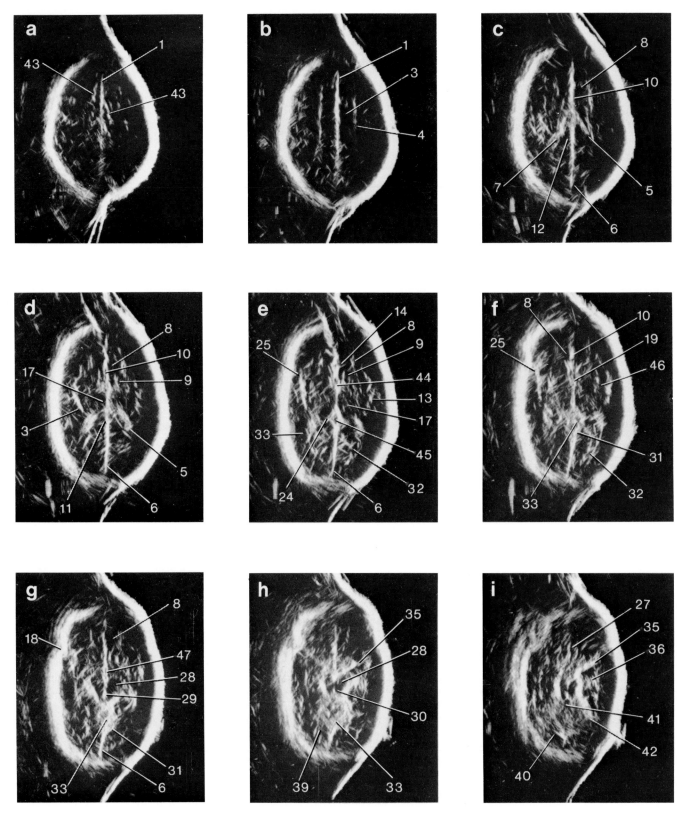

Fig. 1.2 Transverse sections of the brain tilted 20° from the true horizontal plane. Letters refer to plane of section shown on diagram on opposite page.

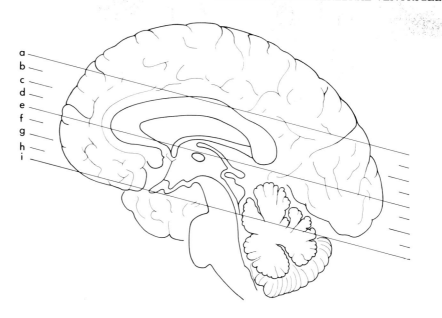

1. Falx cerebri anterior to corpus callosum
2. Body of corpus callosum
3. Body of lateral ventricle
4. Lateral wall of body of lateral ventricle
5. Atrium of lateral ventricle
6. Falx cerebri posterior to corpus callosum
7. Choroid plexus
8. Anterior horn of lateral ventricle
9. Head of caudate nucleus
10. Septum pellucidum
11. Splenium of corpus callosum
12. Forceps major of corpus callosum
13. Sulcus of insula
14. Genu of corpus callosum
15. Cavum of septum pellucidum
16. Body of fornix
17. Thalamus
18. Fronto-parietal suture
19. Foramen of Monro
20. Third ventricle
21. External capsule
22. Lateral aspect of lentiform nucleus
23. Posterior part of hippocampus
24. Choroid fissure
25. Lateral cerebral fissure
26. Olfactory sulcus
27. Straight gyrus
28. Cerebral peduncle
29. Cerebral aqueduct
30. Colliculus
31. Tentorium cerebelli
32. Posterior horn of lateral ventricle
33. Low-level echo area (see text)
34. Vermis of cerebellum
35. Lesser wing of sphenoid bone
36. Temporal lobe
37. Body of sphenoid bone
38. Pons
39. Occipital lobe
40. Cerebellum
41. Fourth ventricle
42. Petrous temporal bone
43. Cingulate sulcus
44. Stylus of septum pellucidum
45. Great cerebral vein
46. Inferior horn of lateral ventricle
47. Optic recess of third ventricle
48. Crus cerebri
49. Falx above the corpus callosum
50. Column of fornix
51. Stem of lateral sulcus★
52. Sulcus of corpus callosum

★ The term 'lateral sulcus' is synonymous with Sylvian fissure.

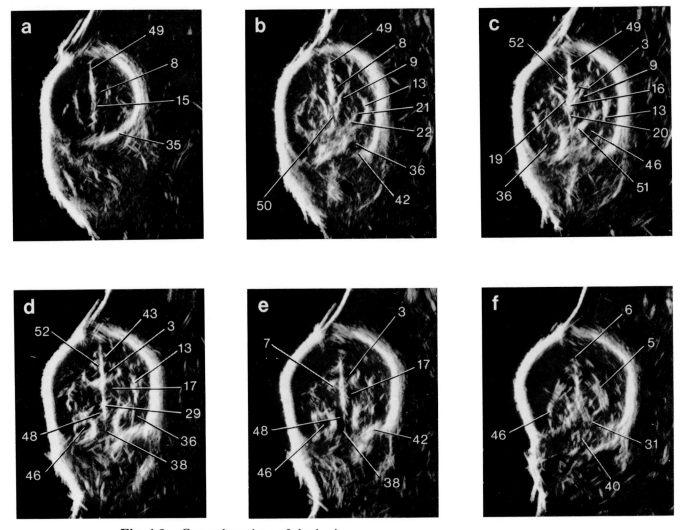

Fig. 1.3 Coronal sections of the brain.

Letters refer to plane of section shown on diagram on opposite page.

THE BRAIN AND CEREBRAL VENTRICLES

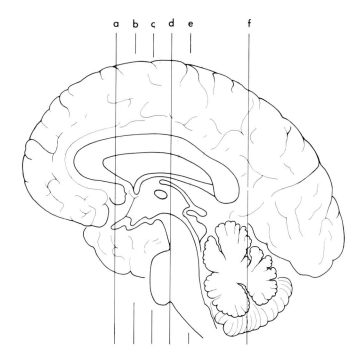

1. Falx cerebri anterior to corpus callosum
2. Body of corpus callosum
3. Body of lateral ventricle
4. Lateral wall of body of lateral ventricle
5. Atrium of lateral ventricle
6. Falx cerebri posterior to corpus callosum
7. Choroid plexus
8. Anterior horn of lateral ventricle
9. Head of caudate nucleus
10. Septum pellucidum
11. Splenium of corpus callosum
12. Forceps major of corpus callosum
13. Sulcus of insula
14. Genu of corpus callosum
15. Cavum of septum pellucidum
16. Body of fornix
17. Thalamus
18. Fronto-parietal suture
19. Foramen of Monro
20. Third ventricle
21. External capsule
22. Lateral aspect of lentiform nucleus
23. Posterior part of hippocampus
24. Choroid fissure
25. Lateral cerebral fissure
26. Olfactory sulcus
27. Straight gyrus
28. Cerebral peduncle
29. Cerebral aqueduct
30. Colliculus
31. Tentorium cerebelli
32. Posterior horn of lateral ventricle
33. Low-level echo area (see text)
34. Vermis of cerebellum
35. Lesser wing of sphenoid bone
36. Temporal lobe
37. Body of sphenoid bone
38. Pons
39. Occipital lobe
40. Cerebellum
41. Fourth ventricle
42. Petrous temporal bone
43. Cingulate sulcus
44. Stylus of septum pellucidum
45. Great cerebral vein
46. Inferior horn of lateral ventricle
47. Optic recess of third ventricle
48. Crus cerebri
49. Falx above the corpus callosum
50. Column of fornix
51. Stem of lateral sulcus*
52. Sulcus of corpus callosum

*The term 'lateral sulcus' is synonymous with Sylvian fissure.

THE CEREBRAL VENTRICLES

There are four intercommunicating *cerebral ventricles*, two lateral, and a third and fourth which lie in the median plane. They are lined by ciliated epithelium and contain cerebrospinal fluid. The choroid plexus invaginates the ependyma into the body of the lateral ventricle on its medial wall and extends forwards as far as the interventricular foramen where it is continuous with the choroid plexus of the other side. Posteriorly it is carried round into the posterior and inferior horns. The cerebrospinal fluid is produced by the choroid plexus, mainly in the lateral ventricles. It drains into the fourth ventricle and passes through the medial and lateral foramen into the subarachnoid space.

The *lateral ventricles* are irregular cavities in the lower medial parts of the cerebral hemispheres, one on each side of the median plane. They are separated by a median vertical septum—the septum pellucidum—but communicate indirectly via the third ventricle through the interventricular foramen of Monro. Each ventricle consists of a central part and three horns: anterior, posterior and inferior. The body and central part extends from the foramen to the splenium of the corpus callosum. It is a curved cavity, triangular in section with a roof, floor and medial wall. The anterior horn passes forwards and downwards into the frontal lobe; in coronal section it is a triangular slit in shape. The posterior horn curves backwards and medially into the occipital lobe. The inferior horn, which is the largest of the three, traverses the temporal lobe curving round the thalamus, passing first backwards and laterally and down and then forwards to within 2.5 cm of the apex of the temporal lobe.

The *third ventricle* is a median cleft between the two thalami. Posteriorly it communicates with the fourth ventricle through the aqueduct, and anteriorly with the lateral ventricles through the interventricular foramen. It has a roof, a floor, an anterior and posterior boundary and two lateral walls. The interventricular foramen is at the junction of the roof with the anterior and lateral walls; in the adult it is crescentic in shape though in the embryo it is relatively larger and circular. The floor of the third ventricle is prolonged down into the infundibulum as the infundibular recess.

The *fourth ventricle* is a lozenge-shaped space in front of the cerebellum. It is continuous inferiorly with the central canal of the medulla oblongata; its superior angle is continuous with the aqueduct of the mid-brain (aqueduct of Sylvius) which is continuous with the third ventricle. It has lateral recesses, dorsal median and dorsal lateral recesses; and a roof, lateral boundaries and a floor.

In the fetus the ventricular system can be seen and measured from the third month. At this stage the lateral ventricles are relatively large and the lateral walls of the bodies are clearly seen. The lateral ventricular width and internal biparietal diameter can be measured to obtain the *lateral ventricular ratio* (LVR). The LVR is the ratio of the width of the body of the lateral ventricle to half the internal biparietal diameter. In early pregnancy it is as high as 0.50; at term or in the neonate the normal range is 0.24 to 0.3. (Garrett & Kossoff 1977).

In the neonate and until the anterior fontanelle closes high quality images of the brain and ventricular system may be obtained by direct contact scanning using the fontanelle as an acoustic window. Conventional static B scanners can be used for the examination but it is simpler and quicker to use small mechanical or phase-steered sector scanners. Similarly small high-frequency linear arrays can be employed (Pape et al 1979). These mobile real-time systems also have the advantage that they may be taken to the ward for the examination, with minimal disturbance of the sick infant.

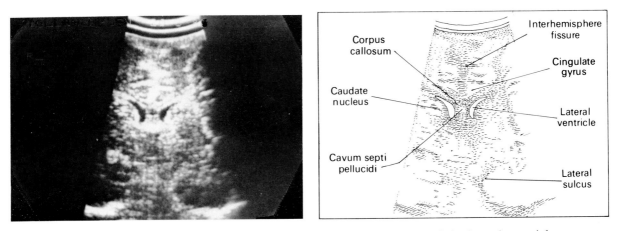

Fig. 1.4 Coronal section at the level of the anterior horns of the lateral ventricles. Contact B scan.

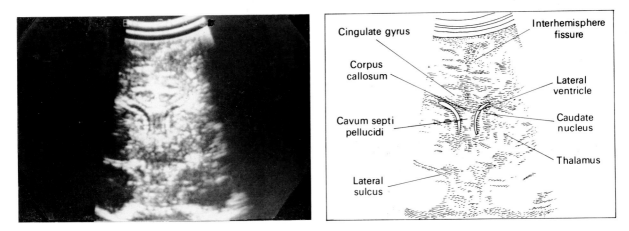

Fig. 1.5 Coronal section through the caudate and thalamic nuclei. Contact B scan.

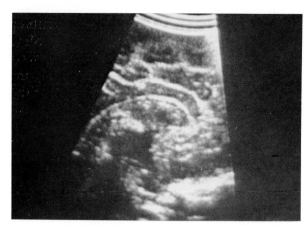

Fig. 1.6 Midline sagittal section.
Contact B scan.

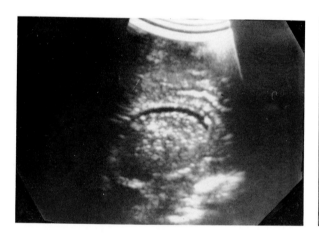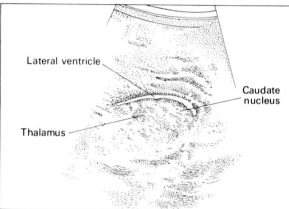

Fig. 1.7 Parasagittal section through the body and anterior horn of the lateral ventricle and the thalamic nucleus.
Contact B scan.

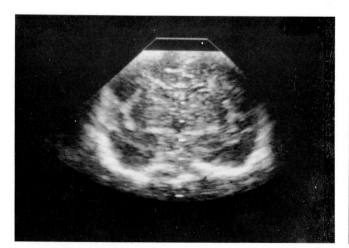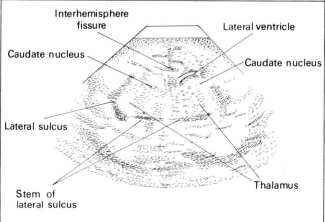

Fig. 1.8 Coronal section at the level of the anterior horns of the lateral ventricles.
Real time sector scan.

THE BRAIN AND CEREBRAL VENTRICLES 11

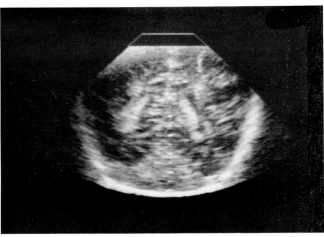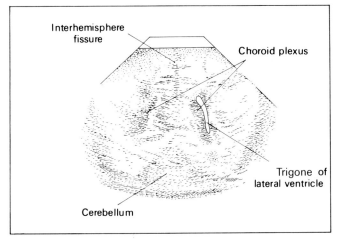

Fig. 1.9 Coronal section through the posterior horns of the lateral ventricles. Real time sector scan.

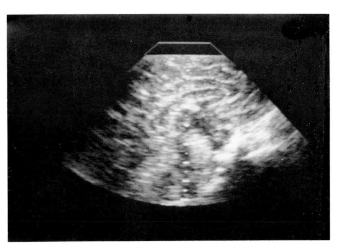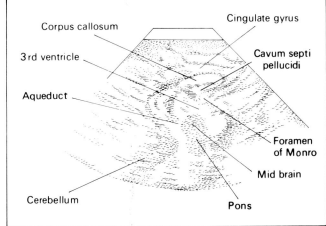

Fig. 1.10 Midline sagittal section. Real time sector scan.

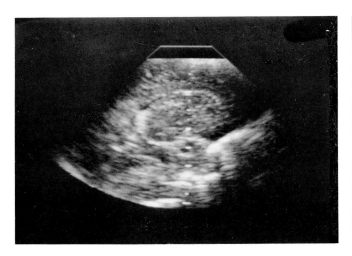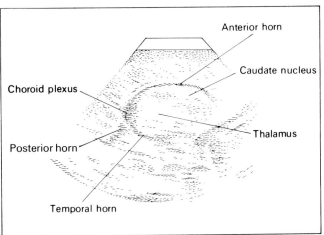

Fig. 1.11 Parasagittal section through the lateral ventricle and thalamic nucleus. Real time sector scan.

2 THE EYE AND ORBIT
Malcolm LeMay

Ultrasonic scanning of the eye was first described by Mundt & Hughes in 1956.

In examination of the eye and orbit it is necessary for sound to penetrate less than 6 cm into the tissues to image all structures from the eyelids to the orbital apex. Focusing of the probe is necessary if maximum resolution is to be obtained. Probes designed for use in ophthalmology are often focused at a distance corresponding to the posterior wall of the eye in an attempt to maximise resolution at this level.

The anatomical structures are mostly small and imaging of what is largely a hollow organ is concerned with detail in and around the walls of the eye. Higher frequencies of ultrasound can therefore be used. The loss of power at depth is not a problem and the improved resolution is a positive advantage in imaging small structures. In the eye resolution of structures smaller than 200 μm is therefore possible. This is an improvement over the value of 500 μm given by Oksala. For these reasons a frequency of 8 MHz or 10 MHz is typically used for the eye. Frequencies from 5 MHz to 20 MHz are often available for orbital and anterior segment imaging respectively (Coleman et al 1969).

Ultrasonic echography of the eye can be performed with the probe in direct contact with the eyelid or cornea, or with a water-bath as acoustic coupling between the probe and the subject. When the direct contact method is used the structures of the anterior segment tend to be lost in the near field of the ultrasonic probe. This is because the probe is refractory to echoes returning after a short interval as it is still vibrating following the 'main bang' of the ultrasonic pulse. There are therefore considerable advantages in using some type of water-bath to enable the probe to be held a few centimetres from the eye, and to facilitate free movement of the probe in compound scanning. If the examination is done with the eyes closed, the cornea is not visible as it tends to be inseparable from the images of the eyelids. It is therefore normal practice to perform ultrasonic examination of the eye with the eye open. The water-bath consists of a polythene sheet with a central aperture for the eye. The polythene is stuck to the face around the eye using a double-sided adhesive disc. The cornea is in direct contact with the liquid of the water bath, which should be sterile and isotonic. With practice the water-bath becomes easy to erect and is well tolerated by patients. Irrigation or injection quality saline or compound sodium lactate solution is used. This liquid should be as near body temperature as possible to standardise the velocity of sound and to minimise disturbance due to thermal currents within the eye. Because high resolution of small structures is required there appears to be an additional advantage to examination with the eye open: if the sound beam passes through the eyelids the thickness of the lids causes sufficient scattering of sound to degrade the final resolution of the image (Coleman et al 1977).

Scans in the transverse orbital plane are the rule in ophthalmology and compound scanning is used or the lateral walls of the globe are not well demonstrated (Coleman et al 1977). The eye is protected by the bony orbit in the other meridia and even in the sagittal orbital plane the bone limits examination to a sector or linear scan. If it is necessary to scan in more than one plane a repeat scan with the eye in a different position of gaze is indicated. Coronal scans of the orbit can be used (Fig. 2.14) but these are constructed by special electronic techniques (Restorie & Wright 1977).

Ophthalmic examination is usually concerned with high resolution of static structures and the compound scan is produced with the eye immobile during the few seconds necessary to complete each scan. Real-time scanning is useful when abnormalities of the vitreous are being explored as these are best seen on eye movement (McLeod et al 1977). Real-time scanning techniques currently used produce a sector scan. The sector scans produced in this fashion show movements of the retina and vitreous, demonstrating fibrosis and points of adhesion between these structures. (The equipment currently in use at Moorfields Eye Hospital, London, produces a real-time linear scan.) However overall resolution tends to be less satisfactory where detailed visualisation of small structures is concerned and so both static and dynamic scanning have an important place in ophthalmic ultrasound.

The scans in the following sections are all compound scans produced during the routine examination of hospital patients at an ultrasonography clinic. Most of these scans are produced with a 10 MHz probe and the water bath technique. The equipment used is the Sonometrics 100 Ophthalmoscan. The equipment is fully described elsewhere (Coleman et al 1969).

14 ULTRASONIC SECTIONAL ANATOMY

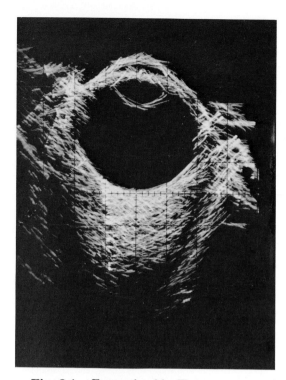 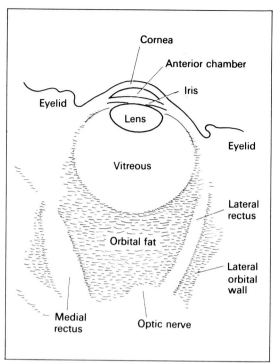

Fig. 2.1 Eye and orbit. Transverse scan in the plane of the orbits. Left eye.

The overall appearance of the eye and orbit are seen. The normal axial length is 24 mm. The velocity of sound is some 10% higher through the lens and the lens is therefore a little foreshortened in a B scan. The medial and lateral rectus muscles are demonstrated. The apex of the orbit is not well visualised. The optic nerve is only demonstrated posteriorly and lies superior to the plane of examination.

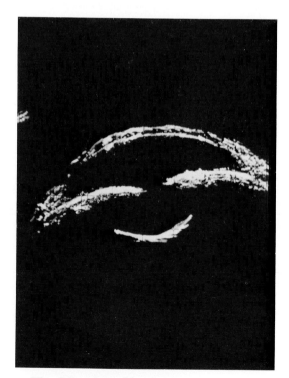 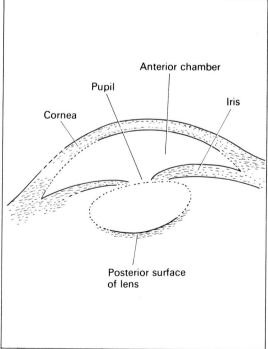

Fig. 2.2 Anterior segment.

The cornea is about 0.6 mm thick centrally and up to 1 mm thick at the periphery. The anterior chamber is bounded by the cornea anteriorly and by the iris and the anterior surface of the lens posteriorly. The shape and depth of the anterior chamber vary, a shallow anterior chamber being formed by an anteriorly more convex lens/iris diaphragm. The shape of the lens will vary slightly with accommodation. The structures of the anterior chamber angle are complex and are not separately demonstrated. The angle of the anterior chamber between the iris and the cornea can be demonstrated and may become closed in some types of acute and chronic glaucoma. In this situation the lens/iris diaphragm appears anteriorly inserted with respect to the external limbus. In this illustration the anterior surface of the lens is not demonstrated. This is partly due to the convex shape and consequent divergence of reflected sound. The anterior lens surface is however easily demonstrated in an eye with a dilated pupil.

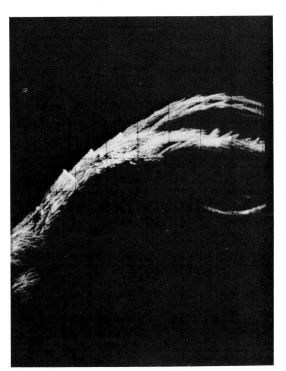 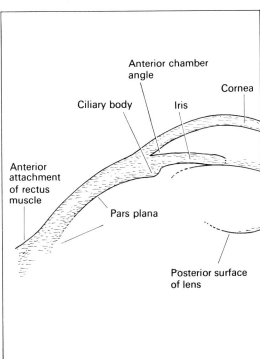

Fig. 2.3 Lateral part of anterior segment.

The area of transition of cornea to sclera is shown. The iris is the anterior part of the uveal layer of the eye and this merges into the structures known as the ciliary body. The ciliary body is posteriorly continuous with the choroid. The angle where the steeper curve of the cornea changes to the curve of the eye is known as the limbus. This is a zone about 1 mm wide. Posterior to this, the ciliary body extends back nasally approximately 6 mm and temporally approximately 7 mm. The ciliary body therefore becomes the choroid some 7 or 8 mm posterior to the limbus as seen externally. The ciliary body itself consists of two parts: posteriorly is the pars plana; anteriorly, the inner surface presents about 70 longitudinal ridges and is thicker than the main mass of the ciliary body. This area is some 2 mm wide and is the pars corona ciliaris. This latter area is referred to as the 'ciliary body' in common usage.

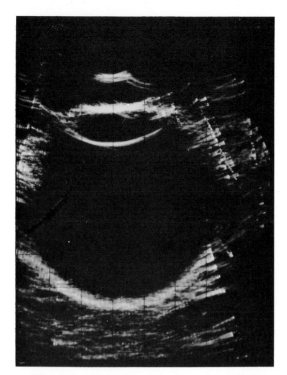 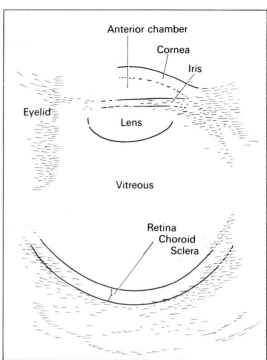

Fig. 2.4 Horizontal section of the globe to show the posterior wall.

In this view the anterior chamber structures are less well seen and the scan has been performed to define the posterior wall. This consists primarily of the sclera which is about 1 mm thick at the posterior pole and becomes thinner further forwards. The globe lies in a thin fibrous capsule (Tenon's capsule) which separates the eye from the orbital contents.

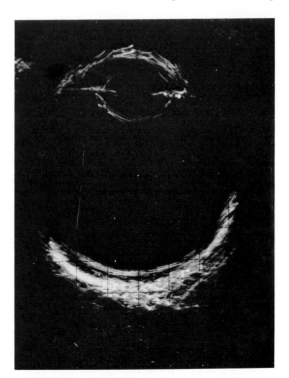 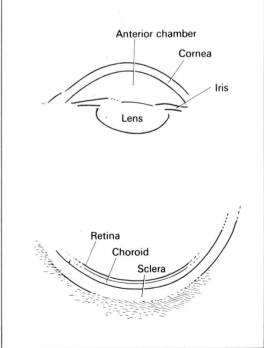

Fig. 2.5 Horizontal section of the globe to show the posterior wall structures.

The retina is some 200 μm thick posteriorly and forms such a well-organised layer that it is often possible to resolve the anterior and posterior surfaces separately. The choroid consists of a less well-organised layer of vessels which show as occasional echoes lying in the space between the retina and the sclera.

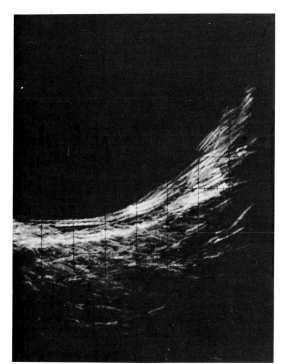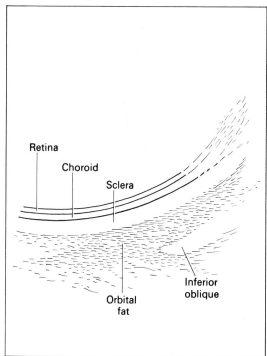

Fig. 2.6 Higher magnification of the posterior wall structures and a demonstration of the inferior oblique muscle.

The interfaces of the retina are seen with the anechoic space of the choroid posteriorly. The choroid is some 200 μm thick in the enucleated eye but is an erectile vascular structure and may be as much as five times thicker than this (1 mm) in life (Coleman et al 1974). The rectus muscles are usually well recognised on ultrasonic scans but the inferior oblique may be seen lying near to or in contact with the globe in the inferotemperal quadrant.

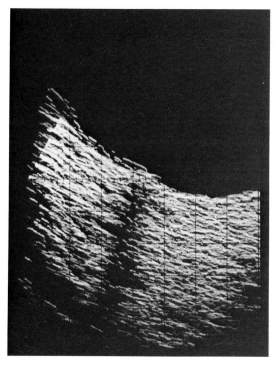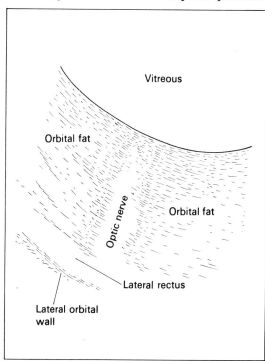

Fig. 2.7 Optic nerve.

The optic nerve is frequently seen obliquely sectioned within the orbital fat. The optic nerve runs a slightly sinuous course from the optic foramen on the medial side of the apex of the orbit to the back of the globe. Its course is variable and depends on the direction of gaze. The nerve is a well-organised structure and because the course of the optic nerve is anteroposterior within the orbit internal echoes are not normally demonstrated. The optic nerve within the orbit is 3 cm long and 3 mm in diameter.

18   ULTRASONIC SECTIONAL ANATOMY

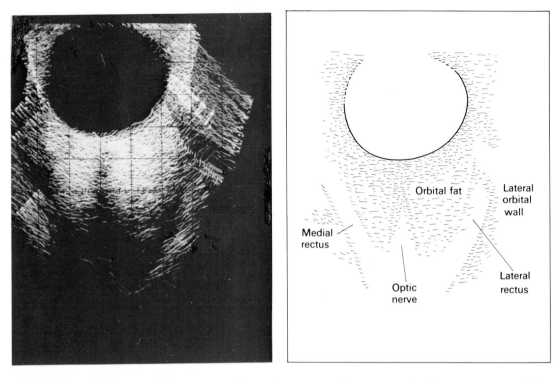

Fig. 2.8A Optic nerve. Movements of the optic nerve are demonstrated with eye movement. The optic nerve with the eye in the primary position.

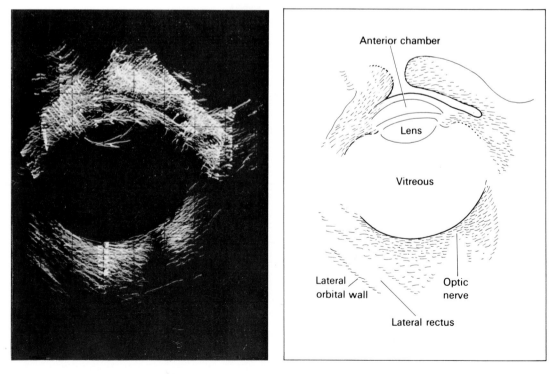

Fig. 2.8B Optic nerve. Movements of the optic nerve are demonstrated with eye movement. Movement of the optic nerve in the orbit with abduction.

THE EYE AND ORBIT 19

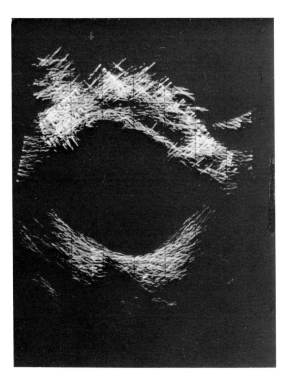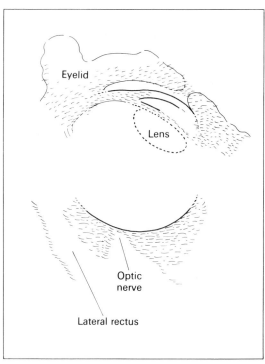

Fig. 2.8C Optic nerve. Movements of the optic nerve are demonstrated with eye movement. Movement of the optic nerve in the orbit with adduction.

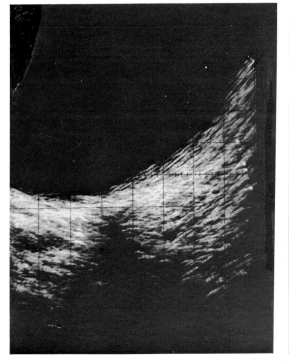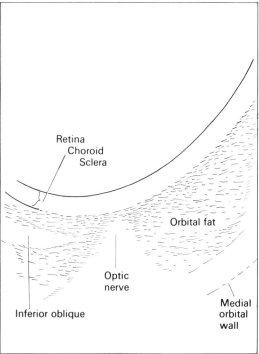

Fig. 2.9 Inferior oblique muscle.

A slightly oblique low transverse section to demonstrate the inferior oblique of the right eye. The inferior oblique is demonstrated lying in the infero-lateral quadrant. The muscle has been sectioned transversely as it lies close to the optic nerve.

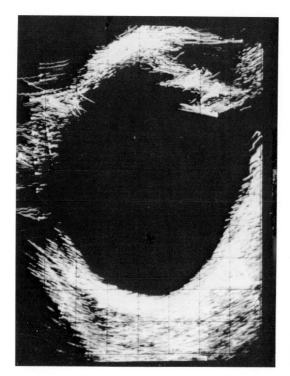 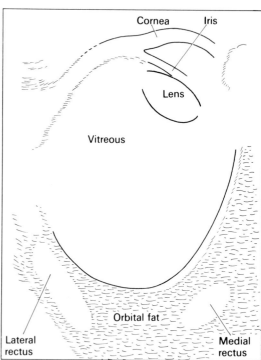

Fig. 2.10 Axial myopia.

An eye with a grossly abnormal axial length. This eye is highly myopic and gave the appearance of proptosis. In this scan the medial and lateral recti are demonstrated in oblique section.

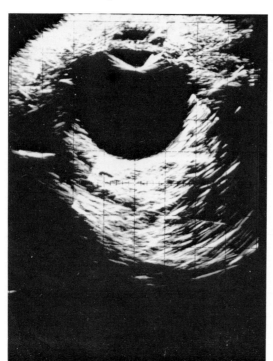 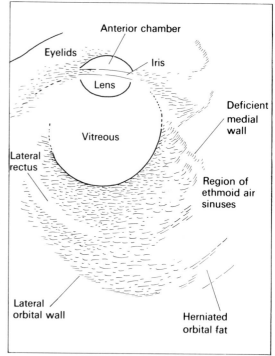

Fig. 2.11 Orbital wall defect.

A horizontal section of the right orbit. The medial orbital wall has been surgically removed in the course of an orbital decompression for relief of endocrine exophthalmos. The orbital fat is increased in amount due to oedema and extends into the area of the deficient ethmoid sinuses. Patients with ophthalmic Graves' disease often show gross enlargement of the extra-ocular muscles posteriorly.

THE EYE AND ORBIT 21

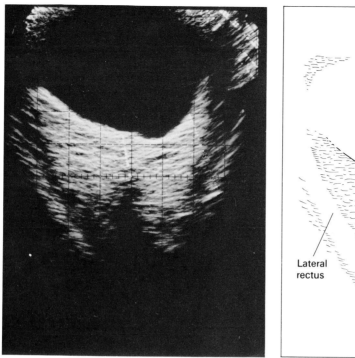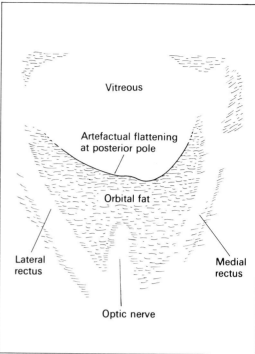

Fig. 2.12 Acoustic artefact from the lens.

Due to a higher velocity of sound through the lens, refraction of sound produces an artefactual shortening of the globe, with flattening of the posterior pole.

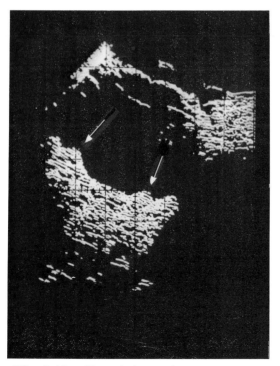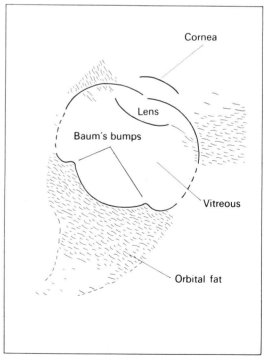

Fig. 2.13 'Baum's bumps'.

This artefact is also due to refraction of sound by the lens. A small section of the posterior wall appears to have a steeper curvature and is displaced slightly anteriorly.

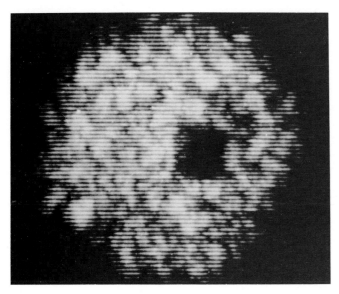 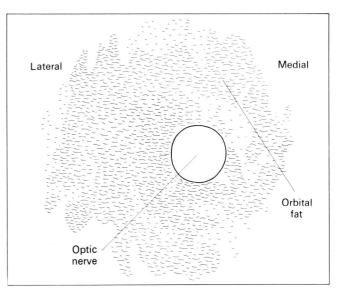

Fig. 2.14 Orbital 'C' scan.

A coronal scan of the orbit produced by the technique of Restorie & Wright (1977). This picture shows a coronal section of the orbital fat pad immediately behind the globe. The optic nerve is clearly seen.

3 THE THYROID AND ADJACENT SOFT TISSUES OF THE NECK

The thyroid may be examined with a conventional static B scanner either by direct contact or by scanning through a water-bath coupled to the neck by acoustic gel or mineral oil. Contact scanning is mechanically difficult and more consistent higher-quality images are usually obtained using the water-bath with the depth of water adjusted so that the thyroid lies in the focus of the transducer beam. Small diameter, high-frequency transducers with a short internal focus should be employed. Recently high resolution real-time contact scanners have been developed for imaging the neck and these produce excellent anatomic and textural detail as well as the added parameter of tissue pulsatility.

The *thyroid* develops as a bud from the floor of the developing pharynx. It descends to reach its definitive location at the level of the hyoid bone and laryngeal cartilage to assume its fully developed configuration. Initially it is connected to the pharynx by the thyroglossal duct which later becomes obliterated. Aberrant tissue may be found along the course of the duct or if the duct fails to obliterate a cyst may form along its tract. Occasionally the thyroid may descend lower than the hyoid bone and thyroid tissue may be found in the superior mediastinum.

The thyroid gland is attached to the lateral and anterior aspect of the trachea by loose connective tissue. Two *lateral lobes* measuring approximately 5 cm x 3 cm x 2 cm are connected inferiorly by an *isthmus* of varying size. An additional *pyramidal lobe* is sometimes present on the superior aspect of the isthmus.

The apex of each lobe extends between the sternohyoid and constrictor muscles of the pharynx. The lateral surface is covered by the infrahyoid muscles, the sternothyroid, sternohyoid and omohyoid muscles. The medial surface rests against the trachea, the constrictor muscles of the pharynx and the oesophagus. The recurrent laryngeal nerve lies behind the lobe of the thyroid inferiorly and is accompanied by the inferior thyroid artery (the minor neurovascular bundle). Higher in the neck the nerve lies between the oesophagus and trachea. The posterior surface is in contact with the paired parathyroid glands, the prevertebral muscles and the carotid sheath which contains the carotid artery, the jugular vein and vagus nerve (the major neurovascular bundle).

Normal thyroid tissue is acoustically homogeneous with moderate density echoes. A firm fibrous capsule, not defined acoustically, surrounds the gland and penetrates the thyroid tissue forming pseudolobules. The connective tissue of the neck is of a higher acoustic density than the thyroid tissue at standard machine settings, but the strap and prevertebral muscles are of a lower density. The trachea is central and, seen anteriorly, it casts a strong acoustic shadow with reverberation artefacts. The carotid artery, internal jugular vein, sternocleidomastoid and prevertebral muscles are routinely outlined: with high quality images the vagus nerve, sympathetic trunk and minor neurovascular bundle may be defined. With present instrumentation, however, the parathyroid glands are not seen unless they are enlarged.

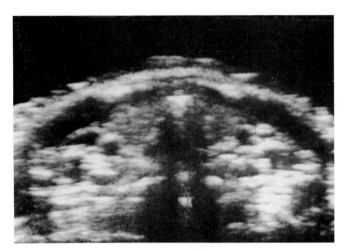 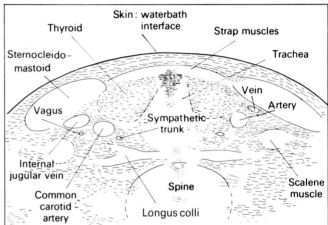

Fig. 3.1 Transverse section at the level of the thyroid. Both lobes are seen on either side of the central trachea, which casts a strong acoustic shadow. Reverberation artefacts are also present. The major vessels are well defined, lying within the deep cervical fascia which is of high density. The minor neurovascular bundle is located medial to the posterior border of the thyroid lobe. The oesophagus is not defined, being obscured by the trachea.

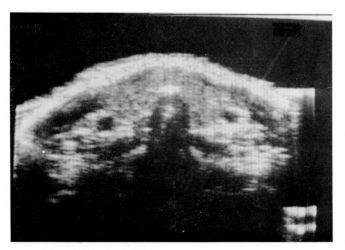 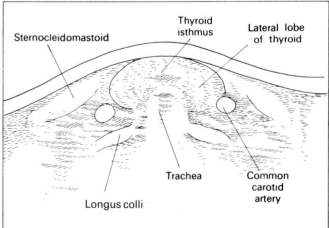

Fig. 3.2 Transverse section at the level of the thyroid isthmus. The isthmus varies considerably in size and is not always easily defined. In this scan it is slightly enlarged.

THE THYRIOD AND ADJACENT SOFT TISSUES OF THE NECK 25

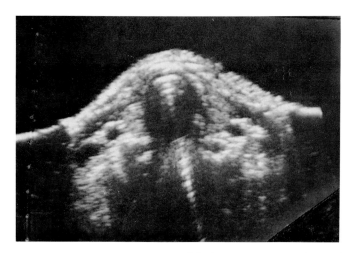 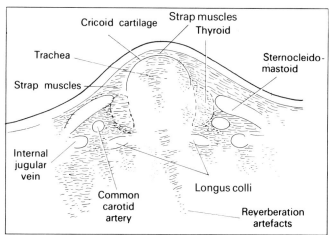

Fig. 3.3 High transverse section taken approximately at the level of the cricoid cartilage. The superior aspect of the lobes of the thyroid is seen at this level. Note the change in the relationship of the internal jugular vein to the common carotid artery as compared with Figure 3.1.

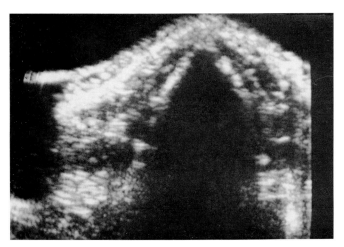 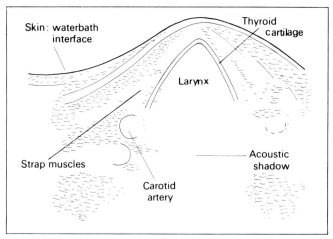

Fig. 3.4 High transverse section at laryngeal level. Only lateral soft tissue detail is obtained—air in the larynx obscures posterior detail.

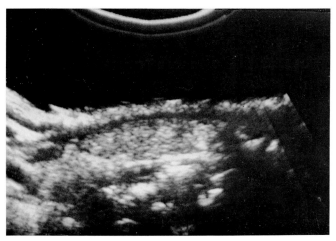 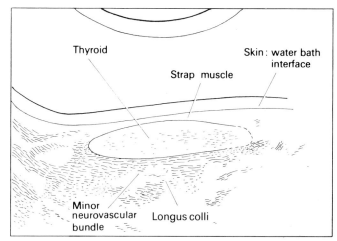

Fig. 3.5A Sagittal section through the medial aspect of the lobe of the thyroid with the minor neurovascular bundle and longus colli muscles seen posteriorly. There are minor density variations in the texture of this thyroid. (Scan reversed with the subject's head to the right of the section.)

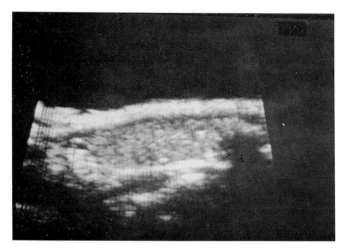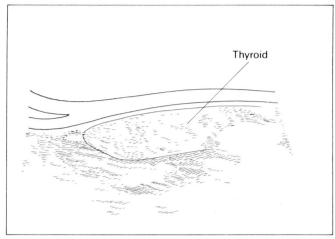

Fig. 3.5B Sagittal section. The thyroid texture is homogeneous and normal. The 'texture' varies with the frequency of the transducer. In this section a 3.5 MHz transducer was employed. (Scan reversed with the subject's head to the right of the section.)

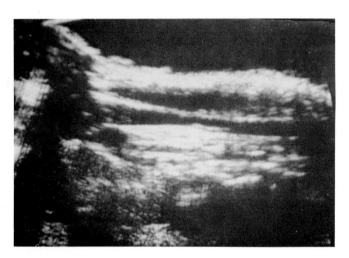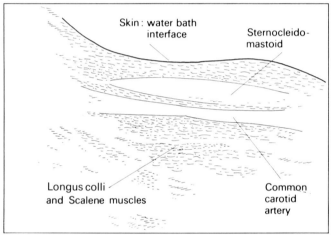

Fig. 3.6 Sagittal section lateral to the thyroid. The common carotid artery is displayed as a tubular structure, and can be identified by its arterial pulsation. (Scan reversed with the subject's head to the right of the section.)

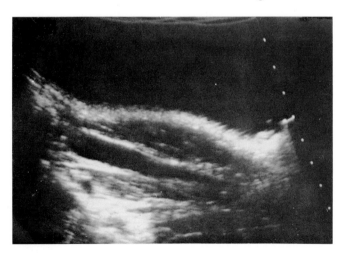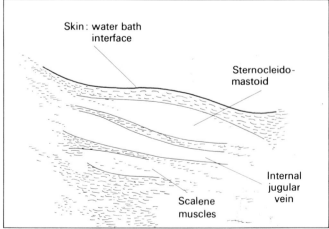

Fig. 3.7 Sagittal scan of the internal jugular vein. This section is lateral to Figure 3.6. The vein varies in size with respiration and may be distended by the Valsalva manoeuvre as demonstrated in Figure 11.51. (Scan reversed with the subject's head to the right of the section.)

4 THE BREAST
William J. Garrett

ECHOGRAPHY OF THE NORMAL BREAST

The *mammary gland* consists of glandular tissue distributed in lobules with connecting fibrous tissue and a variable quantity of fatty tissue in the intervals between the lobules. The subcutaneous tissue which also contains a variable quantity of fat encloses the mammary gland but does not form a distinct capsule. In echograms, the glandular tissue returns higher-level echoes than the fat but the pattern is very variable. The skin covering the normal breast has a smooth contour except at the areola. The nipple and lactiferous ducts immediately deep to the nipple usually give lower-level echoes compared with the glandular tissue within the breast. Apart from during lactation, the lumina of the ducts are not commonly defined. It is not unusual to see a central area of low-level echoes in the normal breast corresponding to an increased fat content at that position.

The echograms shown in the figures are (A) transverse sections and (B) longitudinal sections through the nipple. There is considerable variation in both the anatomical and ultrasonic appearances of the normal breast and this can only be demonstrated by reference to a series of cases.

Editorial note

Breast scanning is usually carried out with specially designed water immersion systems such as have been employed for the scans in this chapter. Multiple sections have to be recorded to examine the whole of the breast. However, high quality scans of localised areas of interest may be obtained with static B scanners using a conventional water-bath technique, or by contact scanning with high-frequency real-time systems.

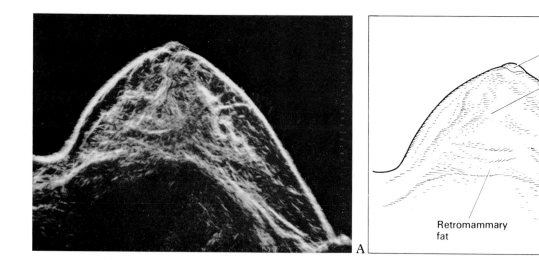

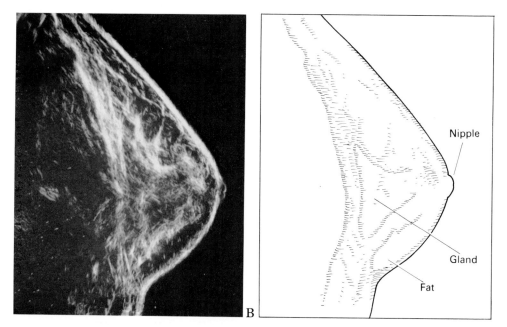

Fig. 4.1 Left breast, para 0, aged 20. The glandular tissue is shown in the transverse section as a triangular grey area embedded in fat. The subcutaneous fatty tissue is 2 to 3 cm thick and the retromammary fatty layer is 1 cm thick. The suspensory ligaments of Cooper are seen passing through the subcutaneous tissue. Transverse (A) and longitudinal (B) sections.

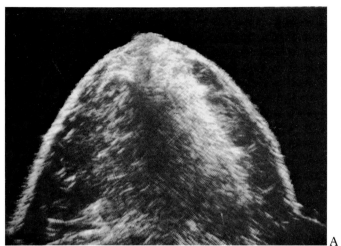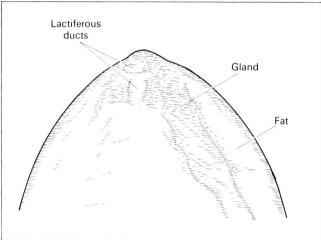
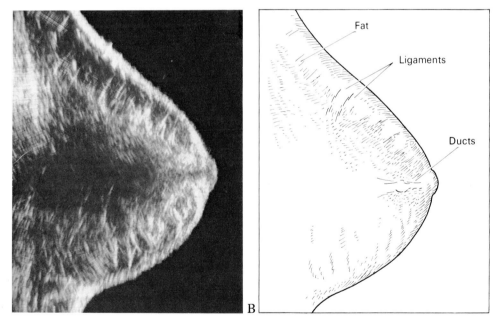

Fig. 4.2 Left breast, para 2, aged 31. A well-marked subcutaneous fatty layer 2 cm thick is outlined, interrupted only by the suspensory ligaments which connect the skin to the glandular tissue. The lactiferous ducts are not seen individually but are recorded as a homogeneous grey band as they pass to the nipple. There is a central area of low-level echoes indicating an increased fat content at this site. This is a normal appearance. Transverse (A) and longitudinal (B) sections.

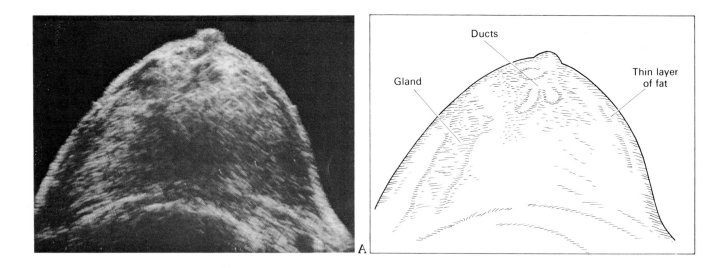

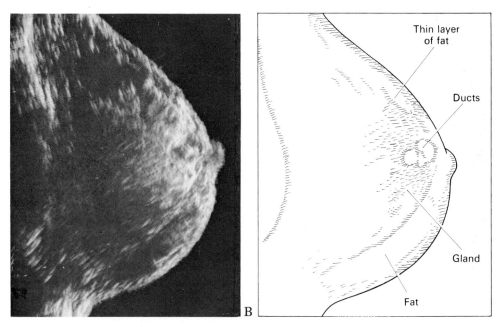

Fig. 4.3 Right breast, para 2, aged 33, 1 day premenstrual. This breast contains very little fat and the subcutaneous layer is barely 5 mm thick in the transverse section at the level of the nipple. The subcutaneous fatty layer is 2.5 cm thick inferiorly and is well demonstrated in the longitudinal sections. This patient complained of considerable premenstrual discomfort and dilated ducts are seen immediately behind the nipple. Transverse (A) and longitudinal (B) sections.

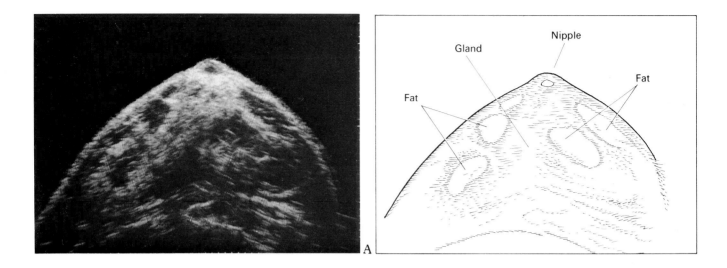

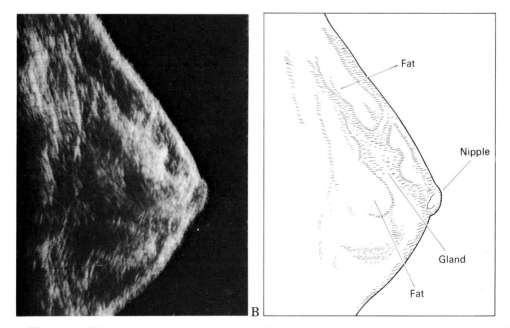

Fig. 4.4 Right breast, para 2, aged 49. The perimenopausal breast shows diminishing glandular tissue. The areas of low-level echoes within the glandular tissue represent interlobular pads of fat. Transverse (A) and longitudinal (B) sections.

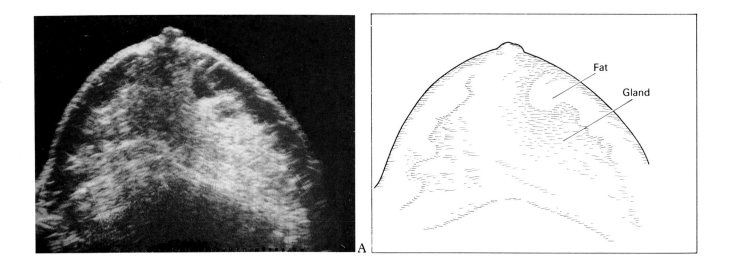

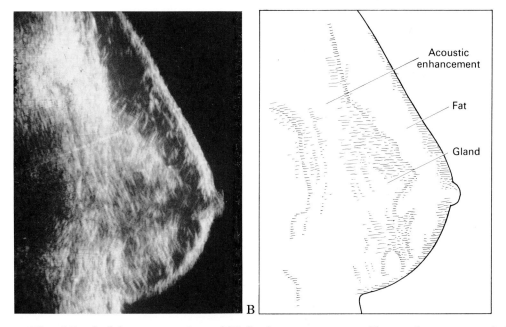

Fig. 4.5 Left breast, para 1, aged 70. In the postmenopausal breast the area occupied by mammary tissue contains mainly fibrous tissue and returns higher level echoes. These echoes and those from the chest wall are much enhanced by the overlying relatively thick fatty layer. This is particularly seen on the lateral side of the breast in the transverse section and in the upper part of the breast in the longitudinal section. Transverse (A) and longitudinal (B) sections.

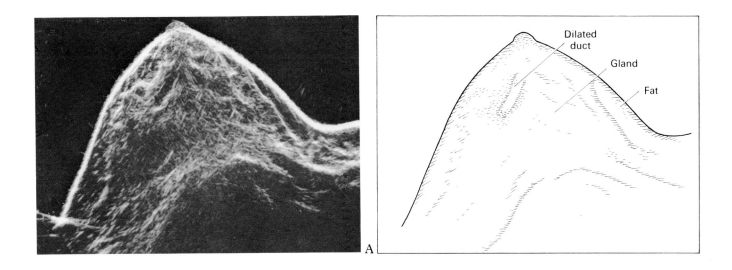

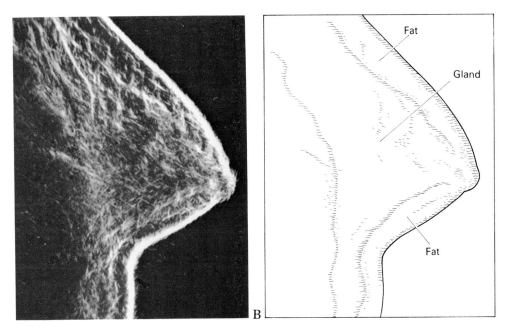

Fig. 4.6 Right breast, para 2, aged 37, 6th day of puerperium. The glandular tissue is well defined in this lactating breast and in the transverse section a dilated duct is seen. Transverse (A) and longitudinal (B) sections.

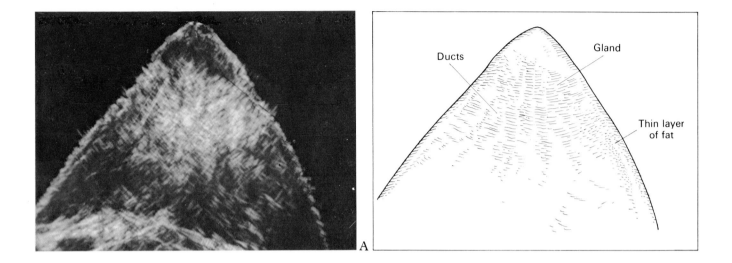

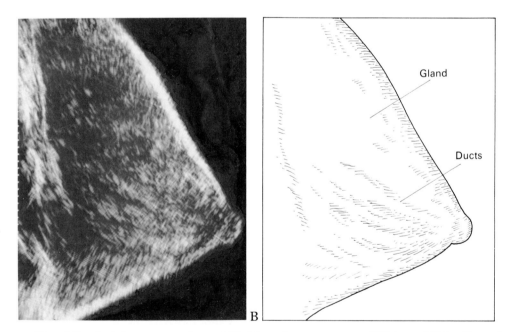

Fig. 4.7 Left breast, para 1, aged 23, 19 weeks post partum. With established lactation the ducts are well defined. The breast in this very thin woman consists largely of glandular tissue with less than 5 mm of overlying subcutaneous fat. Transverse (A) and longitudinal (B) sections.

# 5 THE HEART
James L. Weiss and J. Christine Rodger

Real-time two-dimensional echocardiography displays cardiac anatomy in cross-section. The technique is particularly useful in the investigation of disorders which alter the spatial relationships within the heart. It complements but does not replace M-mode echocardiography which is the preferred method for examining the motion of cardiac structures.

In echocardiography, linear arrays have been superseded by sector scanners. The emphasis in this chapter is therefore on sector scanning. The sector scans reproduced here were recorded with a phased array system (Varian V-3000, Tajik et al 1978) and the linear array scans with a later version (Organon Teknika Echocardiovisor 03) of the system originally developed by Bom (1973).

All scans were recorded on videotape from which Polaroid records were made later. Long-axis sector scans (Figs. 5.1 and 5.2) are presented in the conventional orientation with the chest wall to the top and aortic root to the right as in an M-mode echocardiogram. However, long-axis linear array scans (Figs. 5.11 and 5.12) are presented with the chest wall to the left and the aortic root to the top as in an angiogram. All short-axis images (Figs. 5.3–5.8 and 5.13) are presented as viewed from below with the patient recumbent and the same convention has been followed with the four-chamber views (Figs. 5.9 and 5.10).

RECORDING THE CROSS-SECTIONAL ECHOCARDIOGRAM

The ideal subjects for cross-sectional, as for M-mode echocardiography, are children and thin-chested adults without emphysema. Subjects can be examined supine but access, particularly to the right heart, is often easier with the subject turned to the left. Optimum positioning of the probe depends on the configuration of the chest wall and of the size of the heart and its position within the chest; it must, therefore be determined for each individual with reference to the real-time display.

Sector scanning

The manoeuvrability of the small transducers makes sector scanners particularly suitable for cardiac scanning. Phased arrays and recently developed mechanical systems have the wide (80° or 90°) scanning angle needed for adequate imaging of the adult heart; narrow-angled scanners are of more limited value. Four standard transducer positions, i.e. parasternal, apical, subxiphoid and suprasternal, have been described (Tajik et al 1978, Popp et al 1979).

Parasternal probe position

The probe is positioned at the left sternal edge in the third or fourth interspace. Long-axis images (Figs. 5.1 and 5.2) are obtained by aligning it with the left ventricular long axis. With the probe rotated through 90° from this position, a series of short-axis images (Figs. 5.3–5.8) can be recorded with the level of the cut depending on the tilt of the transducer and on the interspace in which it is positioned.

Apical probe position

The probe is positioned on or internal to the apex beat. Long-axis views are obtained by aligning it with the left ventricular long axis. A short-axis view of the left ventricular apex can be reproduced by directing the transducer at right angles to the apex beat. Upward tilting from this position produces a simultaneous view of all four cardiac chambers.

Subxiphoid probe position

This may provide the best access in emphysematous subjects. The probe is positioned just below the xiphoid process. When it is aligned with the left ventricular long axis, long-axis images are obtained: these are similar to the views recorded from the parasternal position (Figs. 5.1, 5.2) except that the posterolateral rather than the posterior wall of the left ventricle is imaged. When the transducer is rotated through 90° from this position and is then tilted upwards, a four-chamber view is recorded (Figs. 5.9, 5.10, 6.16). With downward tilting, a series of short-axis views can be obtained.

Suprasternal probe position

The probe is positioned in the suprasternal notch. When it is aligned with the trachea, long-axis views of the great vessels can be recorded. Short-axis views are obtained by clockwise rotation of the probe.

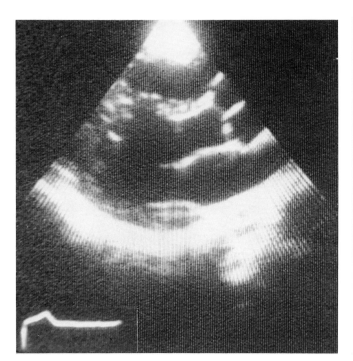 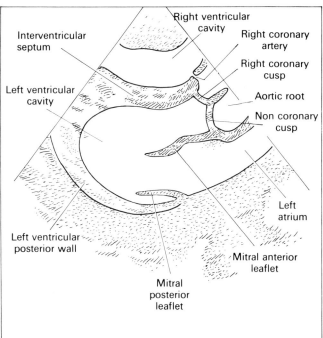

Fig. 5.1 Long-axis sector scan: late diastole (parasternal probe position). The aortic root is related to the right ventricular outflow tract anteriorly and to the left atrium posteriorly. The anterior and posterior walls of the aorta are in continuity below with the interventricular septum and the mitral anterior leaflet, respectively. Though endocardial detail is poor towards the apex, the shape of the left ventricular cavity can be appreciated. The aortic valve is closed: its right and non-coronary cusps are in apposition. The mitral valve is open: both the anterior leaflet and the shorter posterior leaflet can be seen. The origin of the right coronary artery has been tentatively identified; it is not usually seen in long-axis scans and is better visualised in short-axis views of the aortic root (see Fig. 5.7).

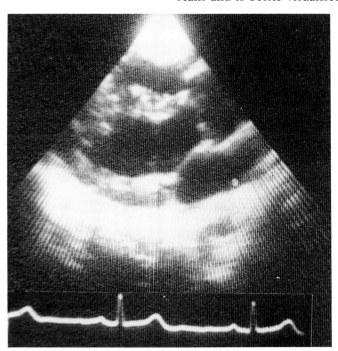 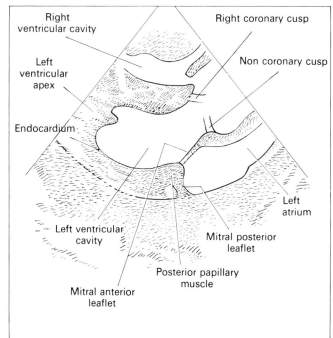

Fig. 5.2 Long-axis scan: end systole (parasternal probe position). Endocardial detail is good and the shape of the left ventricular cavity is clearly defined. Compared with the diastolic scan (Fig. 5.1), the left ventricle has shortened along its major and minor axes, thickening of the interventricular septum is apparent and the posterior papillary muscle can now be seen in close relation to the mitral leaflets. The mitral valve is closed: the anterior leaflet is clearly visible but posterior leaflet detail is obscured by the papillary muscle. Chordae tendineae are not identified. The aortic valve is closing but the right and non-coronary cusps are not in apposition.

38 ULTRASONIC SECTIONAL ANATOMY

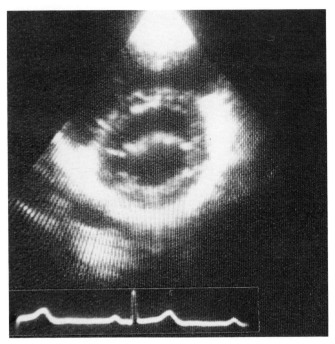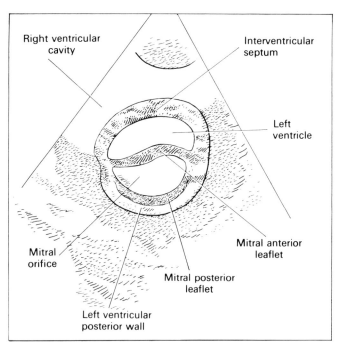

Fig. 5.3 Short-axis scan at mitral valve level: end-diastole (parasternal probe position). Right heart detail is poor: the right ventricle lies in front and to the right of the left ventricle. The entire left ventricular circumference (interventricular septum, anterolateral and posterior walls) has been imaged. The mitral valve is open: the anterior leaflet is bowed forward within the ventricular cavity but the posterior leaflet, which is closely applied to the left ventricular posterior wall, cannot be defined with certainty. The space between the leaflets is the mitral orifice. It should be remembered that the mitral valve is funnel-shaped and that its orifice is maximal at the level of the atrioventricular junction.

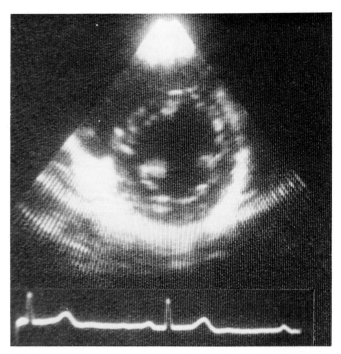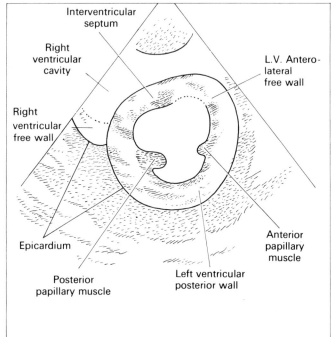

Fig. 5.4 Short-axis scan through left ventricular papillary muscles: end-diastole (parasternal probe position). The right ventricular free wall and within it, the right ventricular cavity, lie in front and to the right of the left ventricle. The entire circumference of the left ventricle has been imaged. The anterior and posterior papillary muscles arise from the anterolateral wall and from the right border of the posterior wall, respectively: their

unattached portions are seen protruding into the left ventricular cavity. It should be appreciated that the papillary muscles are wedge-shaped structures disposed in something approximating the left ventricular long axis (see Fig. 5.2) and that they are not seen in their entirety in the present scan.

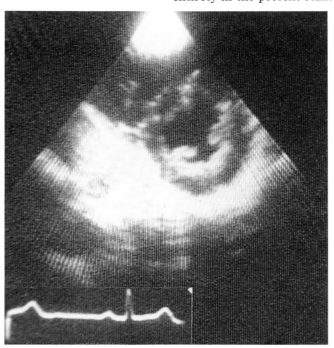 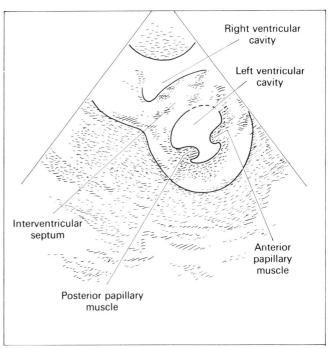

Fig. 5.5 Short-axis scan through the left ventricular papillary muscles: end-systole (parasternal probe position). Compared with the diastolic scan (Fig. 5.4), the left ventricle has shortened along its anteroposterior and transverse axes and thickening of the septum and of the left ventricular posterior and free walls is apparent. The papillary muscle echoes are better defined but the orientation of the papillary muscles is unchanged.

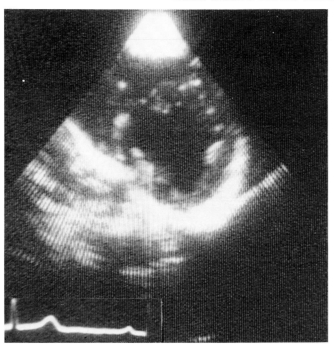 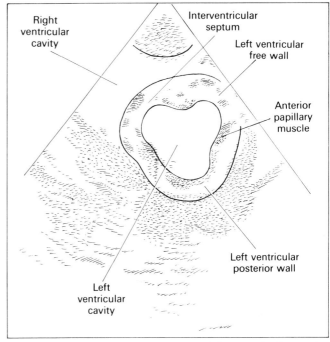

Fig. 5.6 Short-axis scan at apical level: (parasternal probe position). The right ventricular cavity lies in front and to the right of the left ventricle. The entire left ventricular circumference has been imaged: the base of the anterior papillary muscle is seen arising from the anterolateral wall but echoes from the posterior papillary muscle have not been recorded at this level.

40 ULTRASONIC SECTIONAL ANATOMY

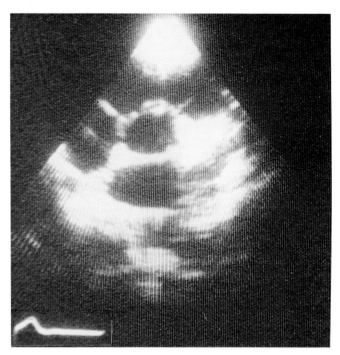 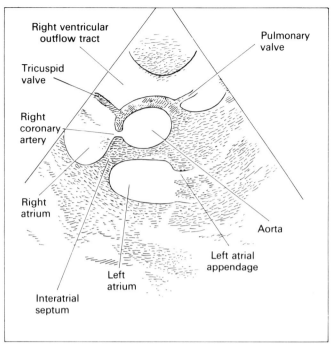

Fig. 5.7 Short-axis scan through aortic root: late diastole (parasternal probe position). The aortic cusps have not been imaged. The right ventricular outflow tract and its continuation, the main pulmonary artery, lie in front and to the right of the aortic root. The pulmonary valve is closed and valve echoes (probably from the posterior cusp) are seen at the origin of the pulmonary artery. The aortic root is related posteriorly to the left atrial appendage and the left atrium and, further to the right, to the interatrial septum. Note that the posterior attachment of the interatrial septum is to the right of the midline and that the plane of the normal interatrial septum is thus at an angle of 45° to the midline. Posteriorly and to the right, the aortic root is related to the right atrium and to the anterior leaflet of the tricuspid valve. The tricuspid valve is partially open and its anterior leaflet is directed towards the right ventricular cavity. Posterior to it, the origin of the right coronary artery is just visible.

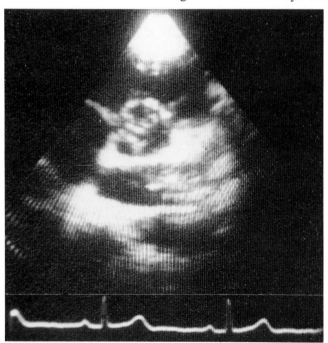 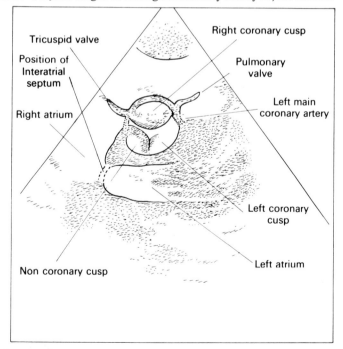

Fig. 5.8 Short-axis scan at aortic valve level: late diastole (parasternal probe position). The tricuspid anterior leaflet, the right ventricular outflow tract and the main pulmonary artery are again clearly defined, but in this scan (cf. Fig. 5.7) only portions of the interatrial

septum can be seen. The three aortic cusps can be identified: the right coronary cusp, which is the largest, lies anterior to both the left and the non-coronary cusps. Because the cusps are thin and move fast, it is usually difficult to image the normal aortic valve in its short axis. This subject had no clinical evidence of aortic valve disease: the disposition of the cusps is normal but it should be noted that the cusp echoes are unusually strong and that diastolic apposition appears incomplete. The origin of the left main coronary artery from the left side of the aortic root behind the pulmonary artery has been tentatively identified.

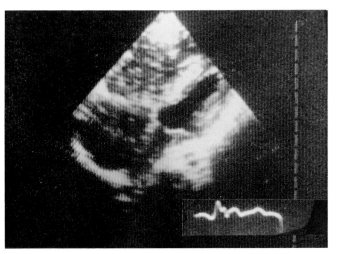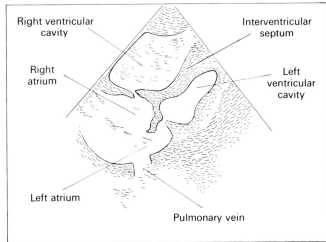

Fig. 5.9 'Four chamber' view with right ventricular injection of saline: end-systole (subxiphoid probe position). The contrast material outlines the right ventricular cavity which lies in front and to the right of the left ventricle. The atria lie behind the ventricles, separated from them by the closed atrioventricular valves and from each other by the interatrial septum. The crux of the heart, the region in which the interventricular and interatrial septa and atrioventricular valves are in continuity, is well displayed: note that the planes of the two septa are not identical and that the right atrioventricular valve is closer to the apex than the left.

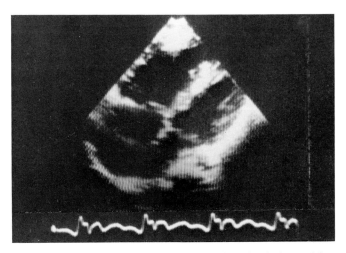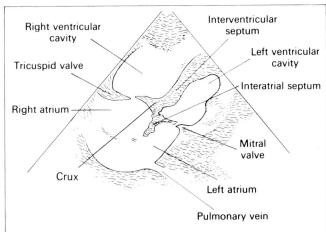

Fig. 5.10 'Four chamber' view without contrast injection, end-systole (subxiphoid probe position). Endocardial detail is good and right ventricular trabeculation can be appreciated. The projections into the left ventricular cavity from the posterolateral wall are probably papillary muscle. The septal and anterior leaflets of the tricuspid valve and both mitral leaflets can be identified.

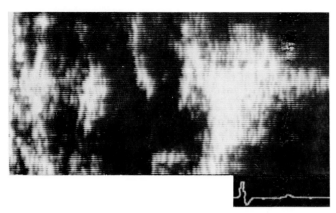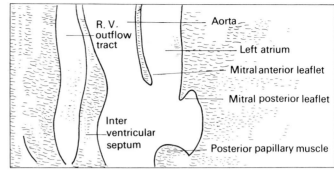

Fig. 5.11 Long-axis linear array scan: late diastole. The aortic root is related to the right ventricular outflow tract anteriorly and to the left atrium posteriorly. The anterior and posterior walls of the aorta are in continuity below with the interventricular septum and the mitral anterior leaflet, respectively. The apex of the left ventricle has not been imaged but the posterior papillary muscle is seen arising low on the posterior wall. The mitral valve is closing: its anterior and shorter posterior leaflets are visible: the subtending chordae tendineae have not been defined.

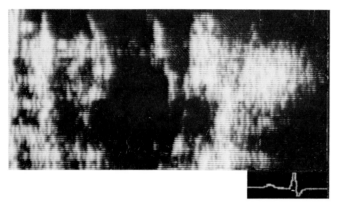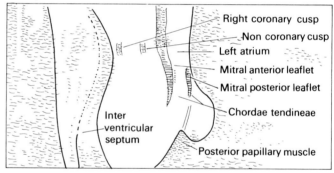

Fig. 5.12 Long-axis linear array scan: mid-systole. The aortic valve is open and its right and non-coronary cusps are visible within the aortic root. The mitral valve is closed: its posterior leaflet has not been defined but the anterior leaflet can be identified and (cf. Fig. 5.11), it is now directed backwards and upwards towards the left atrium. Chordae tendineae to both mitral leaflets are visible; they are aligned in the axis of the posterior papillary muscle.

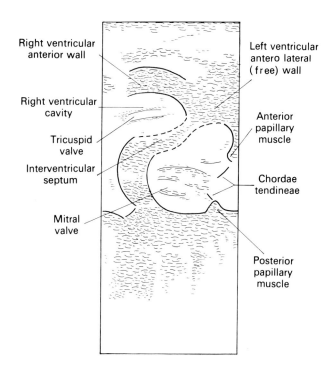

Fig. 5.13 'Short' axis linear array scan through left ventricular papillary muscles: mid-systole. The difficulties of obtaining a true short-axis view with a linear array have been described and comparison with Figure 5.4 indicates that the ventricles have been cut obliquely in this scan. Near-field imaging is superior to the sector scans: the anterior wall of the right ventricle and the left ventricular free wall are well displayed but posterior wall imaging is incomplete. The anterior and posterior papillary muscles are seen arising from the anterolateral wall and from the right border of the posterior wall, respectively. The attachment of the chordae tendineae to the papillary muscle is displayed: the remaining echoes within the left ventricular cavity may be chordal but may be from the mitral leaflets (reflecting the obliquity of the scan).

6 THE ABDOMINAL MUSCLES AND SKELETAL BOUNDARIES

THE ANTERIOR ABDOMINAL WALL

The *anterior abdominal wall* consists of superficial fascia, three layers of muscles and a layer of deep fascia which is continuous with the fascial lining of the abdominal and pelvic cavities. Internal to the deep fascial layer is the parietal peritoneum which is separated from the fascia by a variable amount of extraperitoneal fat.

The *superficial fascia* contains a varied amount of fat, and is usually thickest over the inferior half of the abdomen where it consists of a superficial fatty layer and a deeper membranous layer. The echo density of this fat layer varies with the fat content, but it is usually relatively low-level. The *muscles* consist of three layers which are muscular posteriolaterally and aponeurotic anteromedially. The aponeuroses split to enclose the rectus abdominis muscle and then fuse centrally to form the linea alba.

Details of the anterior abdominal wall can be seen using high-frequency, short internal focus transducers. The appearance however varies considerably depending on the amount of fibro-fatty tissue present and the muscular development. The parietal peritoneum is not defined as a distinct structure as it blends with the deep fascia (Spangen 1976) and extraperitoneal fat layer.

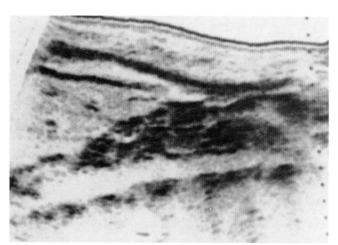 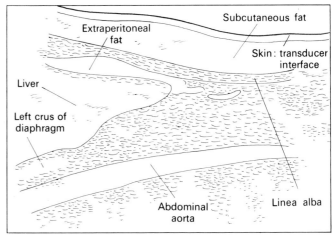

Fig. 6.1 Midline sagittal section in a moderately obese subject with a thick layer of subcutaneous fat. Extraperitoneal fat is seen anterior to the liver in the epigastrium. The linea alba lies posterior to the subcutaneous fat layer and is of high echo density.

THE ABDOMINAL MUSCLES AND SKELETAL BOUNDARIES 45

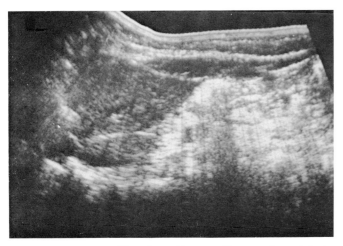 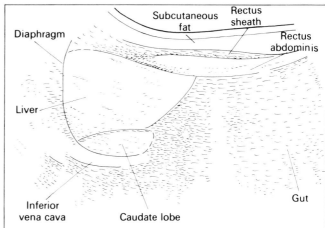

Fig. 6.2 Paramedian section with the rectus abdominis defined behind the subcutaneous fat layer. They are separated by the anterior wall of the rectus sheath.

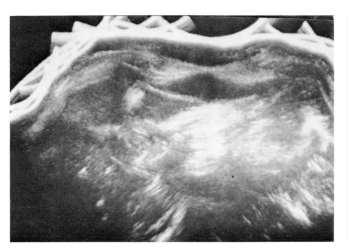 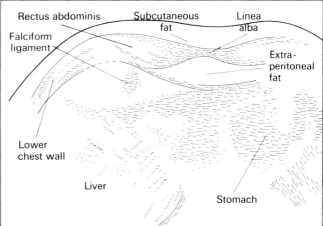

Fig. 6.3 High transverse section in the epigastrium. The extraperitoneal fat and deep fascia are commonly seen at this level behind the rectus muscle which is spreading out for insertion into the lower ribs. The lower part of the falciform ligament, containing the round ligament, is seen as a strong echo in the liver.

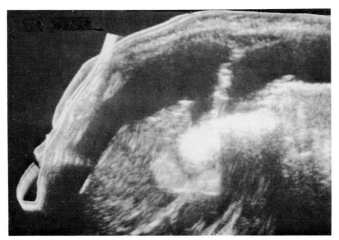 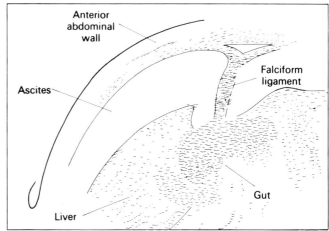

Fig. 6.4 Transverse section in the upper abdomen with cephalad angulation. The falciform ligament is clearly identified in the presence of gross ascites as it passes from the anterior abdominal wall to its attachment to the liver.

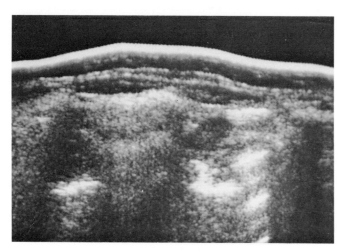 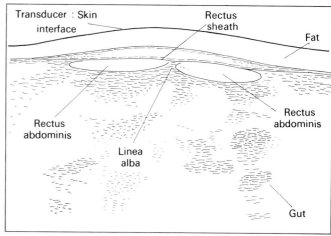

Fig. 6.5 Low subumbilical transverse section. Well-developed rectus muscles are seen behind the subcutaneous fat layer. The linear echo anterior to the rectus sheath may represent the membranous layer of superficial fascia.

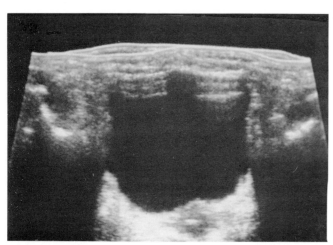 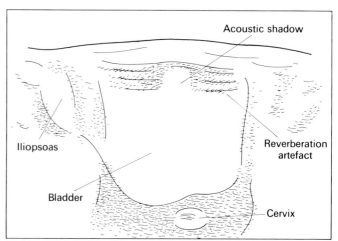

Fig. 6.6 Transverse suprapubic scan. An acoustic shadow is present in the midline caused by thick scar tissue in a central abdominal surgical incision. Reverberation artefacts are seen anteriorly in the bladder.

THE DIAPHRAGM

The muscular fibres of the diaphragm arise from the inferior aspect of the thorax and arch upwards and inwards to form the right and left dome. They fuse centrally as the C-shaped *central tendon* which is slightly depressed by the heart. Posteriorly where the muscular fibres are almost vertical the costophrenic recesses are formed above the diaphragm between the diaphragm and the posterior chest wall.

The *right dome* of the diaphragm which is supported by the liver is relatively accessible to ultrasonic examination. Similarly the right costophrenic recess and right subphrenic spaces can usually be assessed. The *left dome* however is frequently obscured by gastric contents and detail in the left upper quadrant restricted.

The *crura* are thick muscular bundles attached on each side of the aorta to the inferior surface of the lumbar vertebral bodies. The right crus is longer and thicker than the left crus. Their medial sides form the *median arcuate ligament* which arches over the anterior surface of the aorta. Both crura can be seen on ultrasonic sections.

The *three main foramina* of the diaphragm may be demonstrated. The inferior vena cava pierces and fuses with the central tendon of the diaphragm 2 to 3 cm to the right of the midline. It is the highest and most anterior opening. The hiatus for the aorta lies behind the median arcuate ligament anterior to the 12th dorsal vertebra and is the most posterior of the three large apertures. The oesophagus passes obliquely through the diaphragm 2 to 3 cm to the left of the median plane and anterior to the aortic foramina.

The *respiratory movement* of the diaphragm may be studied using conventional B scanners or real-time systems. Its excursion can be documented by recording its position in full inspiration and full expiration on the same image.

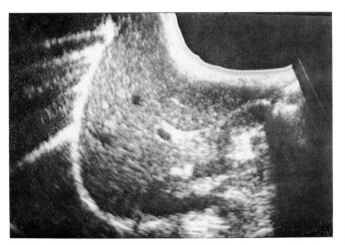 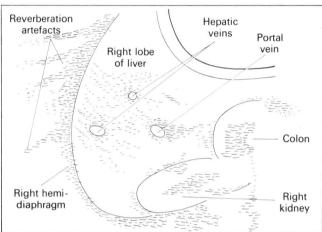

Fig. 6.7 Sagittal section through the liver and the right dome of the diaphragm. Marked reverberation artefacts are present in the lung.

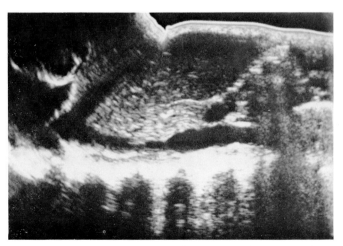 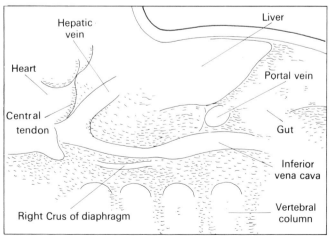

Fig. 6.8 Sagittal section through the opening for the inferior vena cava. This is in the central tendon of the diaphragm 2 to 3 cm to the right of the median plane. The vena cava angles forwards as it passes through the diaphragm anterior to the right crus.

48 ULTRASONIC SECTIONAL ANATOMY

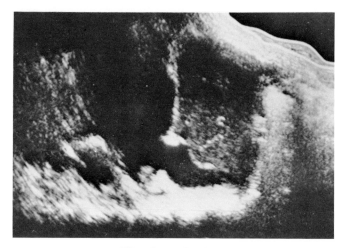 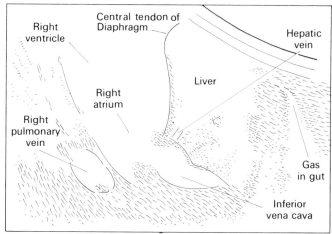

Fig. 6.9 Sagittal section through the opening in the central tendon for the inferior vena cava, with the scan extended above the diaphragm into the right atrium.

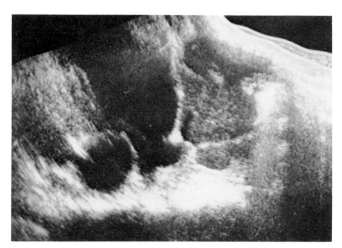 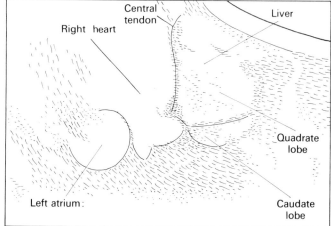

Fig. 6.10 Paramedian section through the central tendon medial to the inferior vena cava.

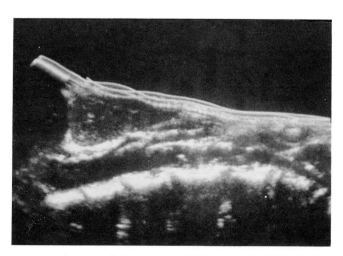 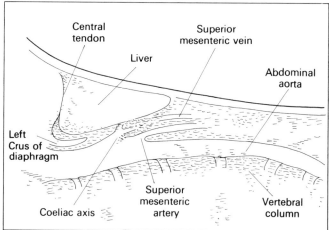

Fig. 6.11 Midline section. The thoracic aorta passes through the diaphragm anterior to the vertebral column. Anteriorly is the right crus of the diaphragm which together with the left crus form the arcuate ligament. This opening in the diaphragm is more posterior than the opening for the inferior vena cava.

THE ABDOMINAL MUSCLES AND SKELETAL BOUNDARIES 49

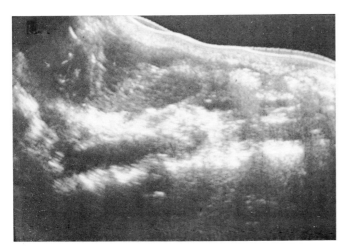 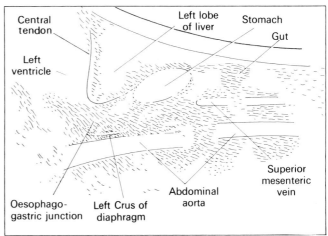

Fig. 6.12 Paramedian section with the opening for the oesophagus anterior to the left crus of the diaphragm and the aorta. The opening runs obliquely through the muscular part of the diaphragm with the central tendon arched over the hiatus.

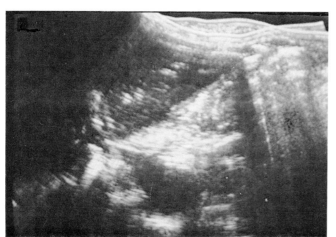 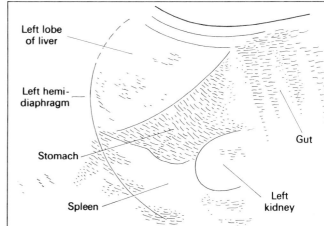

Fig. 6.13 Sagittal section through the left lobe of the liver and the left dome of the diaphragm. Detail of the diaphragm on the left side is frequently restricted by air in the stomach.

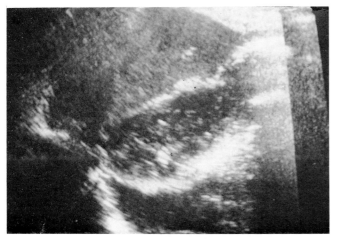 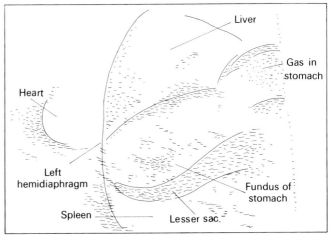

Fig. 6.14 Similar section to Figure 6.13. The diaphragm is better defined as the stomach is distended with fluid. The superior recess of the lesser sac lies behind the stomach, with the splenic recess extending to the hilus of the spleen.

50 ULTRASONIC SECTIONAL ANATOMY

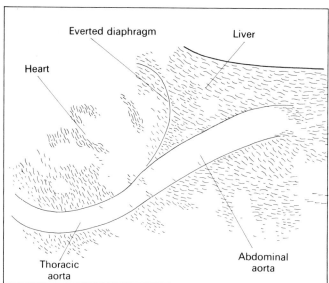

Fig. 6.15 Everted diaphragm in a subject with emphysema and cardiac enlargement. The thoracic aorta is seen behind the heart; it curves anteriorly to pass through the diaphragm at the level of the lower border of the 12th dorsal vertebra. Sagittal section.

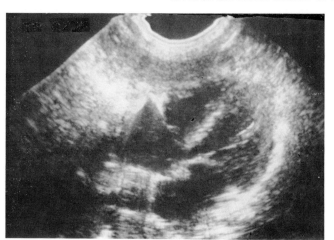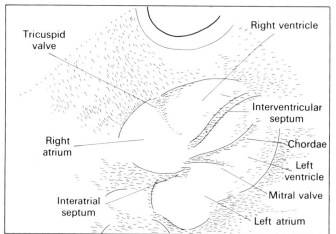

Fig. 6.16 High transverse section angled up through the central tendon to obtain a subxiphisternal four-chamber view of the heart. This scan was taken with a conventional static B scanner and detail is partially blurred by cardiac movement.

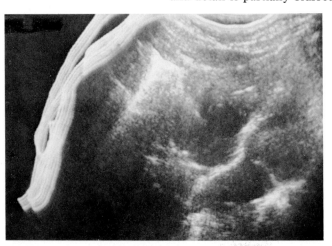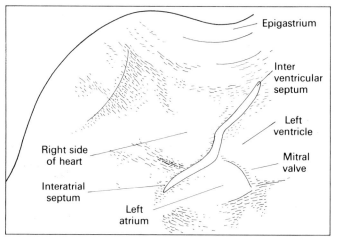

Fig. 6.17 Similar view to Figure 6.16. This subject had a relatively high diaphragm and detail of the intracardiac anatomy is poor.

THE ABDOMINAL MUSCLES AND SKELETAL BOUNDARIES 51

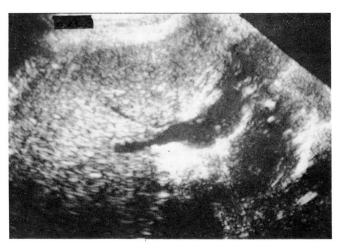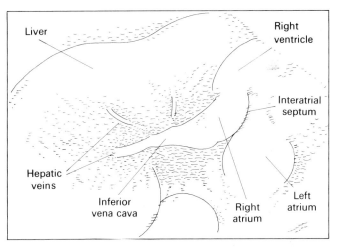

Fig. 6.18 Oblique right subcostal section. The hepatic veins are seen joining the inferior vena cava immediately below the diaphragm. The inferior vena cava passes through its hiatus in the diaphragm directly into the right atrium.

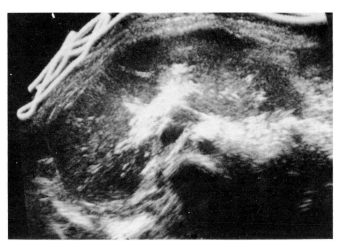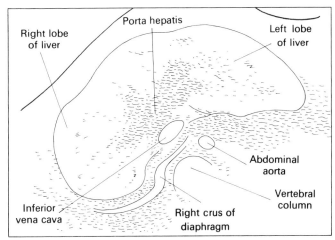

Fig. 6.19 Subxiphisternal transverse section below the diaphragm. The right crus of the diaphragm is seen behind the inferior vena cava and anterior to the abdominal aorta. Laterally it is continuous with the median arcuate ligament which lies anterior to psoas major and psoas minor.

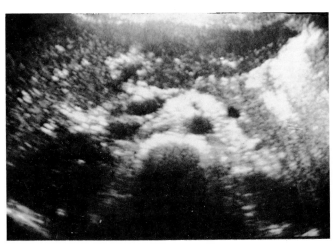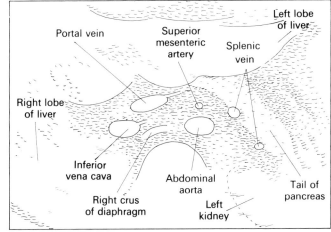

Fig. 6.20 Transverse section at the origin of the portal vein with the right crus of the diaphragm behind the inferior vena cava.

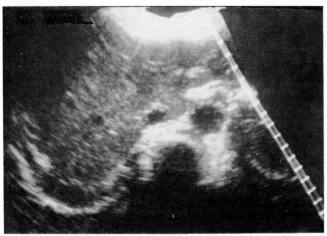 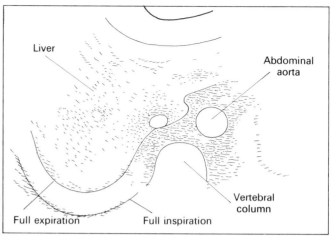

Fig. 6.21 Transverse section demonstrating posterior chest wall movement occurring with respiration. This section was made by recording expiration and inspiration on the one image.

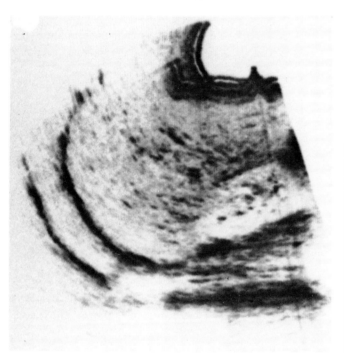 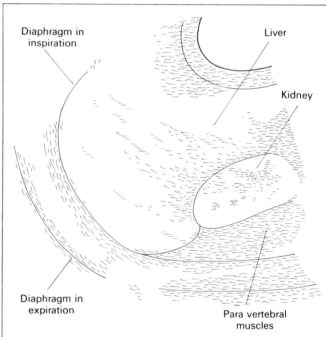

Fig. 6.22 Sagittal section demonstrating the excursion of the right hemidiaphragm with respiration. The two phases of inspiration and expiration have been recorded on the one image.

THE POSTERIOR ABDOMINAL WALL

The three major muscles, the *quadratus lumborum*, *psoas major* and *iliacus* and their surrounding *fascia*, which form the main part of the posterior abdominal wall, can be identified on ultrasonic sections. Psoas minor is not defined separately.

Lumbar vertebral detail is also seen on some sections and represents important anatomic landmarks. The major vessels, the abdominal aorta and inferior vena cava are discussed in the section on visceral vessels. None of the major nerves is defined and normal-sized lymph nodes cannot be differentiated.

THE POSTERIOR ABDOMINAL WALL: SUPINE SECTIONS

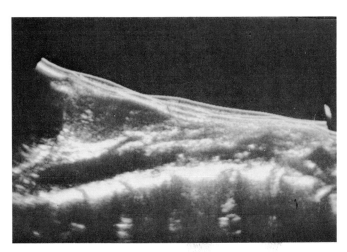 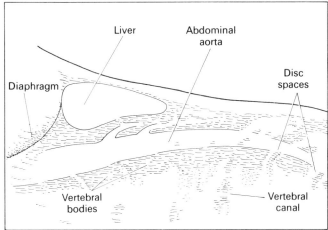

Fig. 6.23 Sagittal section in a slim subject, with the vertebral bodies and disc spaces seen behind the abdominal aorta. The vertebral canal is partially defined.

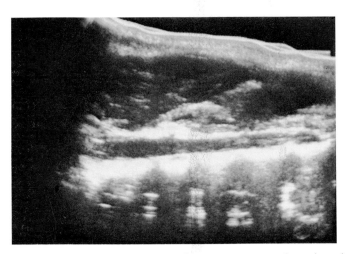 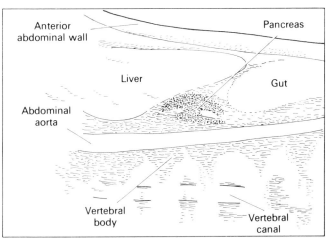

Fig. 6.24 A more penetrated section than 6.23. The vertebral canal is more clearly identified.

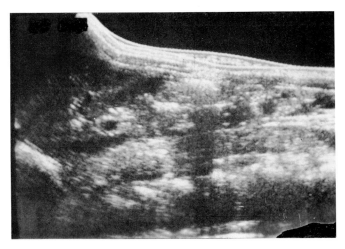 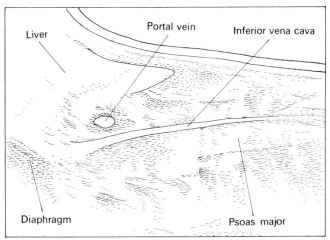

Fig. 6.25 Sagittal section lateral to the vertebral bodies with psoas major seen behind the inferior vena cava. In the cephalic part of the section the muscle mass will also include quadratus lumborum.

54 ULTRASONIC SECTIONAL ANATOMY

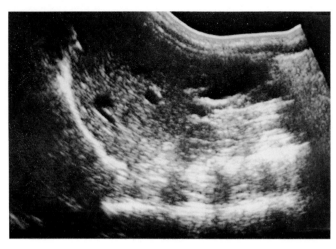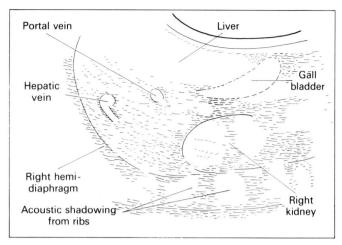

Fig. 6.26 Sagittal section lateral to the right vertical plane. Posteriorly shadowing is seen from the lower ribs.

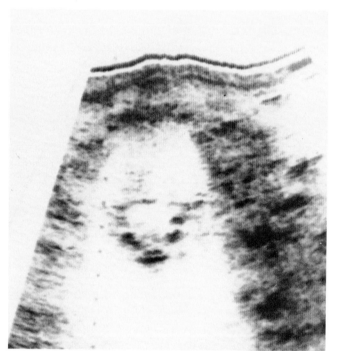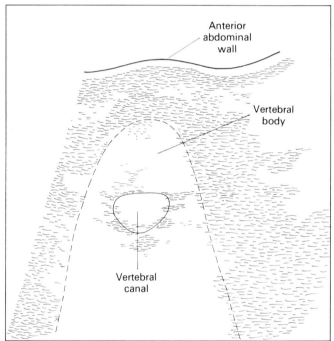

Fig. 6.27 Penetrated transverse section through a lumbar vertebral disc.

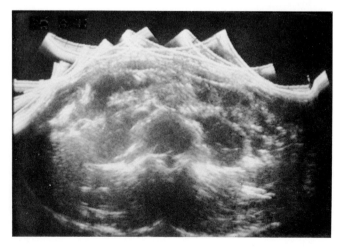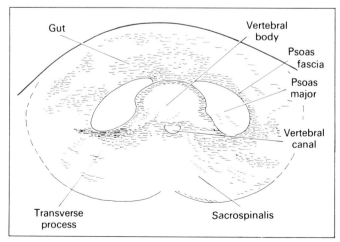

Fig. 6.28 Transverse section above the umbilicus in a subject with well-developed psoas major muscles.

THE ABDOMINAL MUSCLES AND SKELETAL BOUNDARIES 55

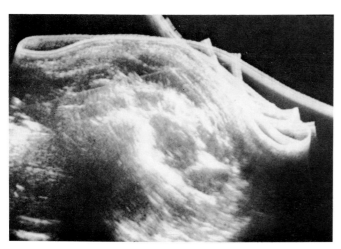 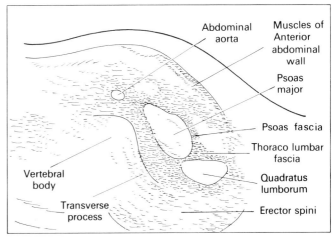

Fig. 6.29 Transverse section demonstrating the relationship of psoas major and quadratus lumborum. The psoas and thoracolumbar fascia are of high acoustic density and are well defined in this scan.

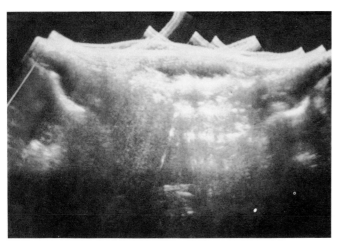 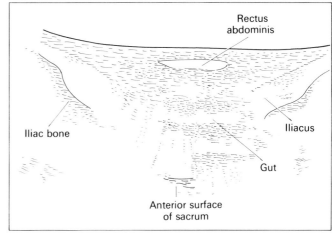

Fig. 6.30 Transverse section through the iliac fossae with iliacus defined laterally. Posterior to the muscle the anterior surface of the ilium is seen forming the floor of the iliac fossa.

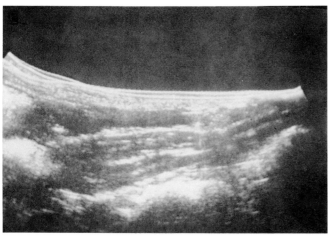 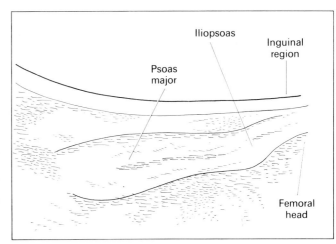

Fig. 6.31 Longitudinal oblique section through the lower abdomen, iliac fossa and inguinal region. The psoas major and iliopsoas are well developed in this subject and are seen passing under the inguinal ligament anterior to the femoral head.

THE POSTERIOR ABDOMINAL WALL: PRONE SECTIONS

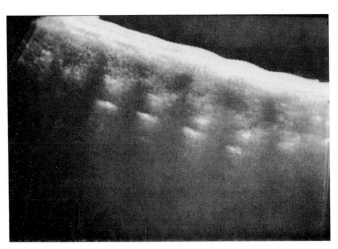 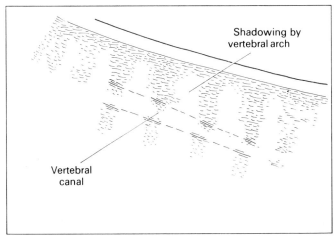

Fig. 6.32 Midline sagittal section outlining the lumbar vertebral canal.

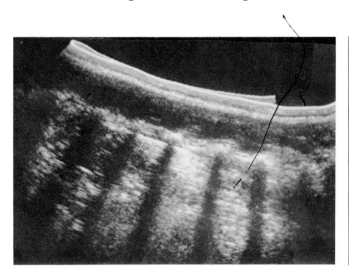 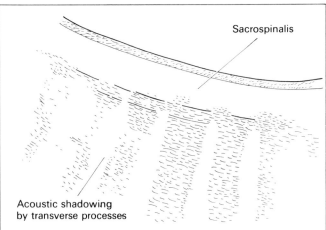

Fig. 6.33 Paramedian section with shadowing from the lumbar transverse processes.

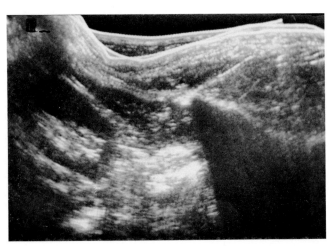 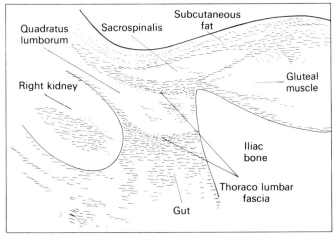

Fig. 6.34 Sagittal section defining the muscle and fascial planes below the lower pole of the kidney.

THE ABDOMINAL MUSCLES AND SKELETAL BOUNDARIES 57

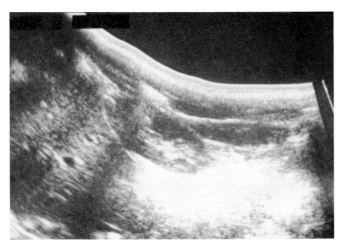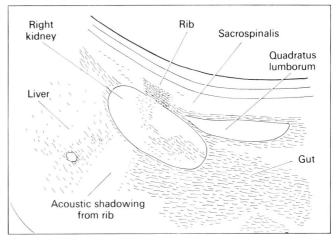

Fig. 6.35 Longitudinal scan through upper lumbar region. Acoustic shadowing is commonly encountered from ribs which may obscure the upper pole of the kidney.

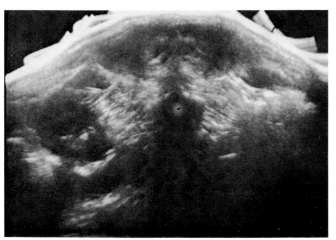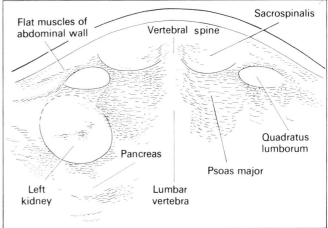

Fig. 6.36 Transverse section through the lumbar region. There is cephalad angulation of the scanning plane.

SKELETAL BOUNDARIES AND MUSCLES OF THE PELVIS

The pelvis is divided into two main parts, the *greater pelvis* formed by the iliac fossae and the *lesser pelvis* which lies below a plane passing from the sacral promontary to the symphysis pubis.

The *lesser pelvis* is bordered posteriorly by the sacrum and coccyx, inferiorly by the pubic bones and symphysis and laterally by the rami of the pubis and ischium surrounding the obturator foramen, and more superiorly by the medial wall of the acetabulum. The posterior lateral wall of the lesser pelvis is divided by the sacrotuberous and sacrospinous ligaments into the sciatic foramina, and the anterolateral obturator foramen is filled by the obturator membrane. Inferiorly the walls end in the boundaries of the inferior aperture of the pelvis or the boundaries of the perineum. This inferior aperture transmits the urethra, the vagina in the female, and the anal canal. Anteriorly it is closed by the urogenital diaphragm and posteriorly by the levator ani muscles.

The walls of the lesser pelvis are lined by muscles, the *piriformis* on the sacrum, *the coccygeus* on the sacrospinous ligaments and the *obturator internus* on the anterolateral walls.

All the walls are covered by a layer of pelvic fascia. The parietal pelvic fascia lines the wall of the pelvic cavity and covers the superior aspect of the levator ani muscles. It is continuous with the visceral pelvic fascia that surrounds the pelvic viscera where these structures lie on or pierce the levator ani. The pelvic peritoneum covers the superior aspect of the pelvic viscera and forms pouches between the viscera. Between the pelvic peritoneum and pelvic fascia there is fatty extraperitoneal tissue containing the blood vessels and hypogastric nerve plexuses.

Anatomic detail may be outlined in the greater pelvis in the slim subject, unless intestinal gas obscures distal information. However to examine the lesser pelvis it is essential that the bladder is distended. The pelvic bones are strongly reflective structures and surface outline detail only is obtained. Soft tissue detail may be clearly displayed, though high-quality sectional detail is dependent on correct swept gain and sensitivity settings. The pelvic connective tissue is of high echo density and the muscle planes, pelvic viscera and larger vessels which are less echogenic can be imaged on selected sections.

The pelvis normally contains small bowel which prevents ultrasonic examination of the lesser pelvis. To overcome this, pelvic examination is carried out with the urinary bladder filled. This is achieved by an oral water load taken 1 to 1½ hr prior to the examination, though occasionally retrograde filling through a catheter may be necessary. The fluid-filled bladder displaces the intestine upwards into the lower abdomen and acts as an acoustic window, permitting detailed imaging of the pelvic viscera and soft tissue planes. It is also possible to examine the bladder walls in detail. The distended bladder acts as a useful anatomic reference point during pelvic examination.

SKELETAL BOUNDARIES

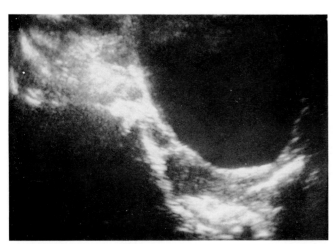 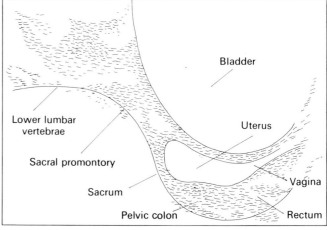

Fig. 6.37 Sagittal section through the posterior superior wall of the pelvis. The lesser pelvis lies below the level of the sacral promontory. The coccyx, which is not defined, is at a level posterior to the upper third of the vagina. Behind the uterus there is a pouch of peritoneum—the rectouterine pouch of Douglas—which is not outlined unless fluid-filled.

THE ABDOMINAL MUSCLES AND SKELETAL BOUNDARIES 59

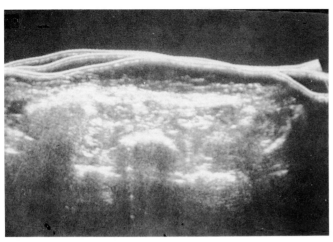 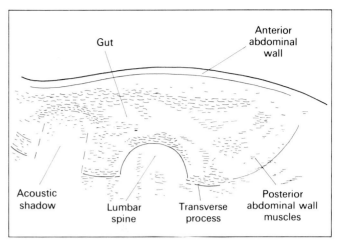

Fig. 6.38 Transverse section below the umbilicus above the level of the iliac crests. The acoustic shadow is from gas in the caecum.

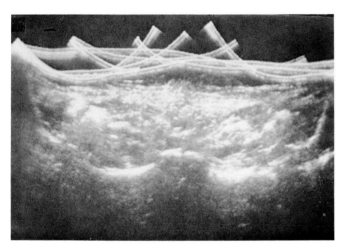 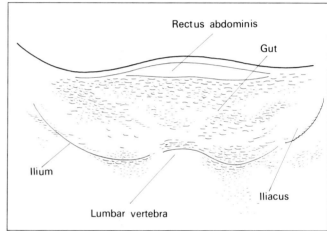

Fig. 6.39 Transverse section below the iliac crests in the intertubercular plane. The section passes through the iliac fossae and is part of the greater pelvis.

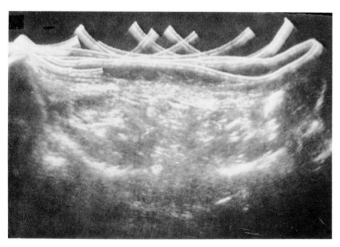 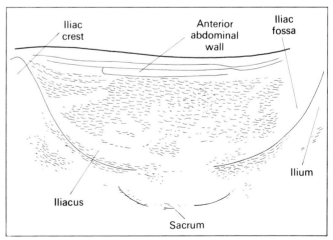

Fig. 6.40 Transverse section through the lower part of the iliac fossa above the anterior superior iliac spine. This section is in the greater pelvis which is deepening, with the sacrum seen posteriorly.

60 ULTRASONIC SECTIONAL ANATOMY

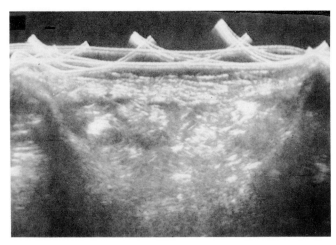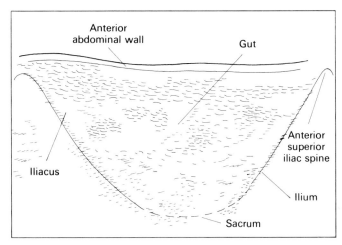

Fig. 6.41 Transverse section at the level of the anterior superior iliac spine. The pelvic cavity is deeper and narrower with the sacral bay posteriorly. This section passes through the greater and lesser pelvis cavities.

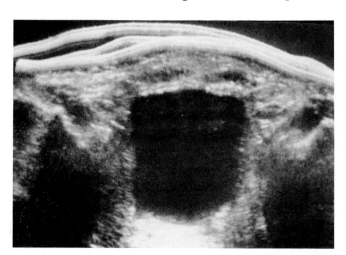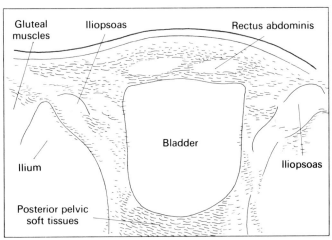

Fig. 6.42 Transverse section between the anterior superior iliac spine and the acetabular region. The sides of the lesser pelvis are vertical, those of the greater pelvis are more inclined to the horizontal plane.

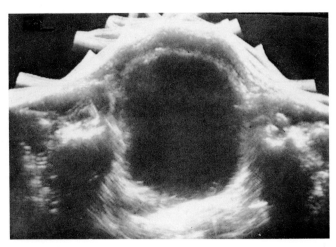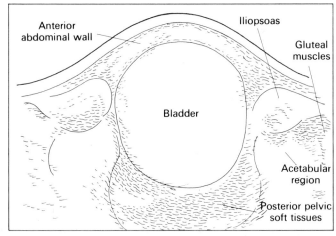

Fig. 6.43 Transverse section above the symphysis pubis through the upper acetabular region. The shape of the bony pelvis has now changed and the sides are vertical. This scan has been extended laterally so that the outline of the pelvic bone is displayed. The femoral nerve sheath is seen as an echogenic area in iliopsoas. (See Fig. 6.45.)

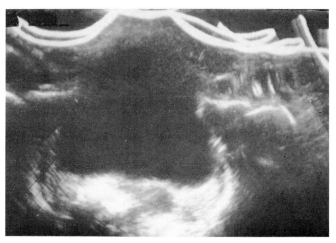
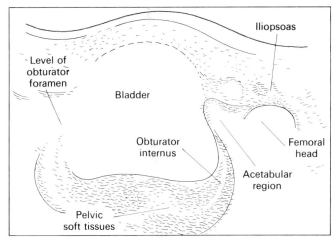

Fig. 6.44 Transverse section through acetabular region and femoral head. The obturator foramen lies at this level forming the anterolateral pelvic wall. The obturator internus is only partially defined posteriorly.

THE PELVIC MUSCLES

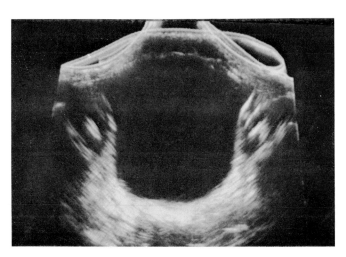
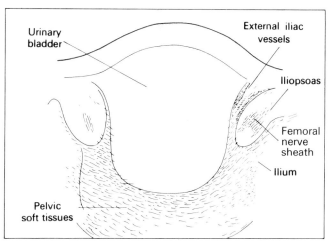

Fig. 6.45 Transverse section with cephalad angulation demonstrating the iliopsoas muscles and the external iliac vessels on their medial aspect. The central high-density echo is commonly seen in the medial aspect of the muscle, it represents the femoral nerve sheath.

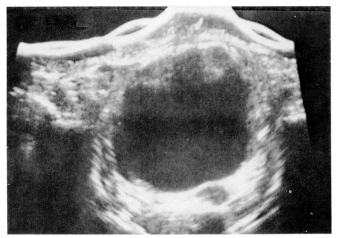
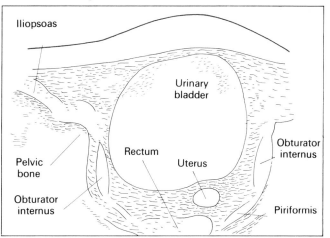

Fig. 6.46 Transverse section through the small pelvis at acetabular level. The obturator internus muscles are seen laterally against the pelvic side wall. They are best defined with caudad angulation of the section. Posteriorly part of piriformis is outlined.

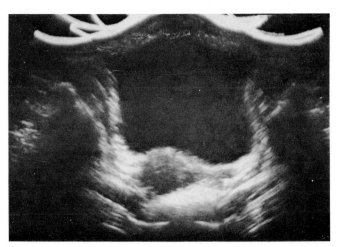
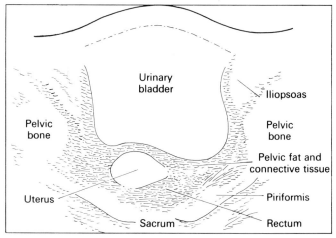

Fig. 6.47 Low transverse section with cephalad angulation. This plane outlines the piriformis muscles posteriorly where they lie anterior to the pelvic surface of the sacrum. Piriformis lies just superior to the coccygeus muscle which forms the lowest part of the posterior wall of the lesser pelvis and is sometimes seen on transverse sections. Note that the pelvic fat and connective tissue is of a higher echo density than the pelvic viscera and muscles. The femoral nerve sheath is seen in the iliopsoas.

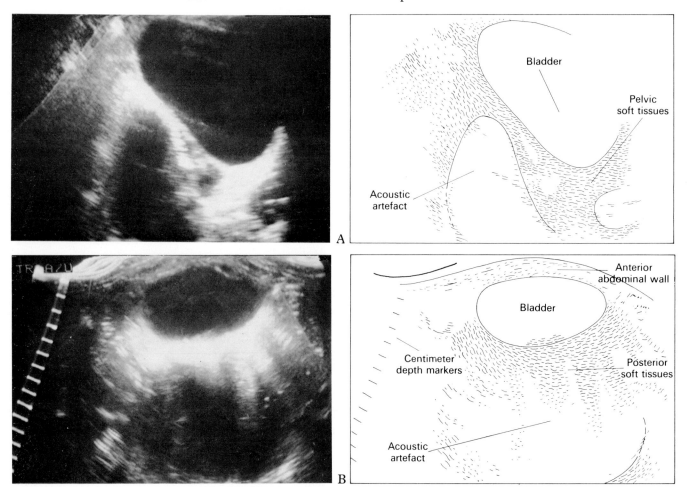

Fig. 6.48 Sagittal (A) and transverse (B) section of an acoustic artefact commonly seen in pelvic images which may be mistaken for a large pelvic mass.

This artefact is produced by the strongly reflective gas behind the bladder causing reverberation deep within the pelvis. In the transverse section reverberation and ringdown phenomena are also present anteriorly. The pseudo-tumour may appear cystic or solid depending on the amount of system noise recorded.

7 UPPER ABDOMINAL VISCERA

THE LIVER, HEPATIC VEINS AND INTRAHEPATIC PORTAL SYSTEM

The *liver*, which is the largest organ in the body, occupies the right hypochondrium, much of the epigastrium and extends into the left hypochondrium. Except in the epigastrium the greater portion of the liver is surrounded by the thoracic cage. With the subject supine the upper border of the superior surface of the liver lies anteriorly at the level of the fourth and fifth rib on the right and the sixth rib on the left. The shape of the normal liver has been shown to vary considerably (Mould 1972).

The liver has a diaphragmatic surface and a posterior inferior or visceral surface. The *falciform ligament* divides the liver topographically but not anatomically or functionally into right and left lobes. The true anatomic left lobe is divided into medial and lateral segments, the medial segment lying to the right of the falciform ligament and the lateral segment representing the classic left lobe. In addition on the visceral surface several fissures and fossae, arranged in the shape of an 'H', demarcate two additional lobes, the caudate lobe and the quadrate lobe. The cross bar of the H is the *porta hepatis*, the inferior half of the vertical fissure contains the *ligamentum teres* and the superior half has the *lesser omentum* attached in its depth with the *ligamentum venosum*. The right margin of the H is represented by the fossae of the gall bladder and the inferior vena cava, and also represent the line of division of the liver into true anatomic right and left lobes. The liver is covered by peritoneum, except for an area on the posterior superior surface adjacent to the inferior vena cava where it is in direct contact with the diaphragm—the bare area of the liver. *Peritoneal reflections* from the anterior abdominal wall, the diaphragm and the abdominal viscera form distinct ligaments which form the boundaries of the subphrenic and subhepatic spaces.

The liver is supplied by the hepatic artery and the portal vein. The *common hepatic artery* arises from the coeliac axis and passes to the liver in the lesser omentum to the left of the common bile duct and anterior to the portal vein. It gives off three major branches—the gastroduodenal artery, the supraduodenal artery and right gastric artery before dividing into right and left ramus. The right ramus passes behind the common hepatic duct and gives off the cystic artery before entering the liver. This classic anatomical pattern is only seen in approximately 60% of subjects.

The *portal vein* carries blood to the liver from the stomach, small intestine, large intestine and pancreas. It is formed from the junction of the superior mesenteric vein and splenic vein behind the neck of the pancreas and passes behind the first part of the duodenum in the lesser omentum to the porta hepatis where it divides into right and left branches. The right branch passes horizontally to the anatomic right lobe, the left branch has a more vertical path to the left lobe.

The venous outflow from the liver is carried by valveless *hepatic veins* which enter the inferior vena cava just below the diaphragm. Beginning with the central vein in the liver lobules the venous blood passes through sublobular and collecting veins into the major right, middle and left hepatic veins. The middle and left usually join to enter the vena cava as one vessel. Several smaller veins from the caudate lobe and other parts of the liver are consistently found.

Sonographic examination of the liver is best performed in a longitudinal direction using an arc-like sweep under the costal margin. Scans are taken in full inspiration. The diaphragm is seen outlining the superior margin of the liver. Sometimes a lateral oblique or erect approach is required if the liver is located at a high level. Transverse views of the liver are limited by the presence of ribs but a fairly large sweep between the costal margin is usually possible and smaller sector scans can be obtained between ribs.

The acoustic texture of the liver is considered to be better appreciated with a black background (white echo format) than with a white background (black echo format). Both hepatic and portal veins can be seen within the liver parenchyma. Hepatic veins branch in a caudad direction and do not have a strong echogenic rim since they are not surrounded by fibrous tissue.

Portal veins tend to branch in a cephalic direction and are surrounded by an echogenic rim of fibrous tissue (Glisson's capsule). Tubular structures seen anterior to the portal veins represent biliary radicles; the right hepatic duct is commonly identified anterior to the right portal vein and in the porta hepatis the common bile duct and hepatic artery lie anterior to the main portal vein.

The size of the lobes of the liver is variable; one well-known variant is the development of Reidel's lobe, where the right lobe is much more extensive than usual. The volume of liver which can be adequately examined depends much on the position of the liver and its relationship to the costal margin. This varies considerably with the stature of the normal subject. A few livers (perhaps 10%) are never seen due to their high position. To obtain a uniform return of echoes from the inferior border of the liver to the superior surface just below the diaphragm requires careful selection of transducer frequency and swept gain settings. The swept gain slope is optimised to reach maximum at the diaphragm; the slope will therefore vary with the section of liver under examination and will vary significantly between different normal subjects.

Figures 7.1 to 7.6 are a series of sagittal sections in the upper abdomen showing the variation in the shape of the liver in subjects of differing physique.

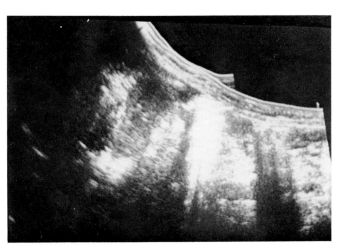 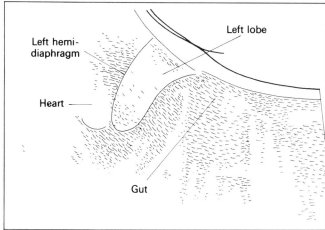

Fig. 7.1 Sagittal section in the left vertical plane. The section is through the lateral segment of the left lobe which is small and high.

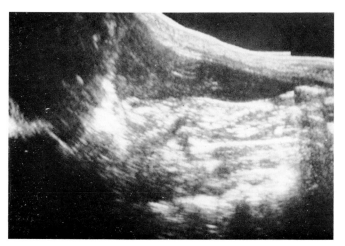 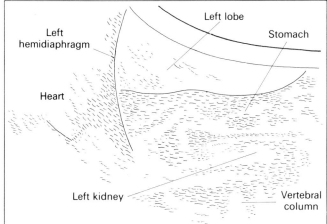

Fig. 7.2 Sagittal section in left vertical plane. The left lobe is long, thin and wedge-shaped, extending well below the costal margin. Small branches of the portal vein are indistinctly defined.

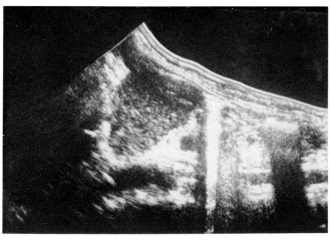 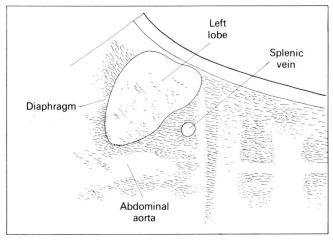

Fig. 7.3 Sagittal section in the plane of the abdominal aorta. The lateral segment of the left lobe is small with a rounded inferior border.

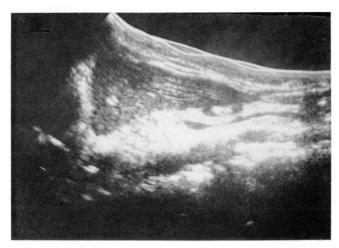 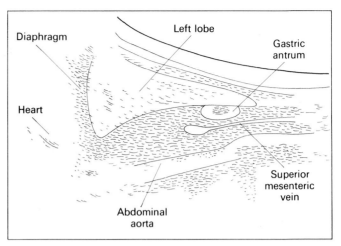

Fig. 7.4 Sagittal section in the plane of the abdominal aorta. The left lobe is long and wedge-shaped. Portal venous radicles are seen below the diaphragm.

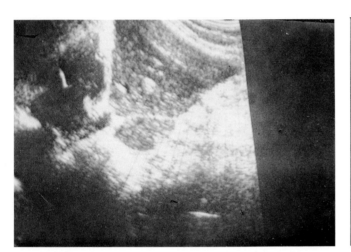 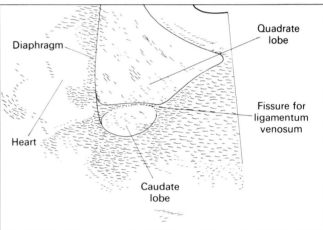

Fig. 7.5 Median section between the abdominal aorta and inferior vena cava through the anatomic medial segment of the left lobe. The fissure for the ligamentum venosum which contains the lesser omentum lies in a horizontal plane separating the caudate and quadrate lobes. Portal venous radicles are represented by high-density echoes from their surrounding hepatobiliary capsule (Glisson's capsule).

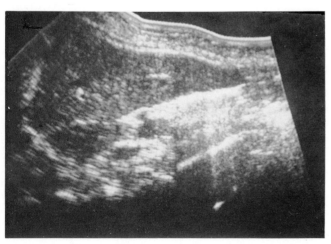 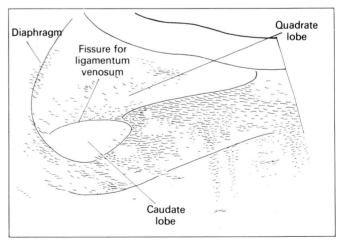

Fig. 7.6 Median section. The liver is long and thin extending low in the epigastrium. The inferior border remains wedge-shaped. Portal veins are seen in the liver parenchyma.

UPPER ABDOMINAL VISCERA 67

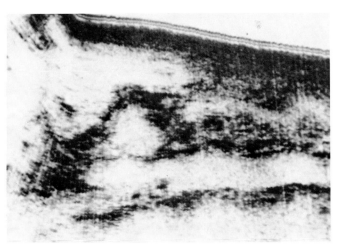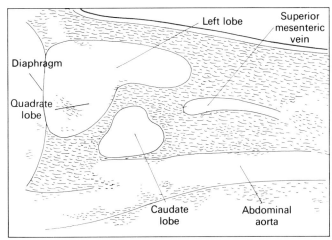

Fig. 7.7 Paramedian section. There is an unusual configuration to the caudate lobe which is separated from the quadrate lobe by a large fascial space. A transverse section of this liver is seen in Figure 7.23.

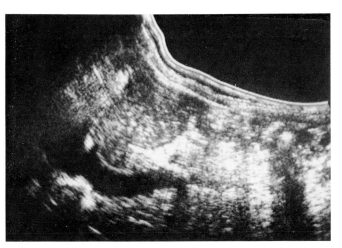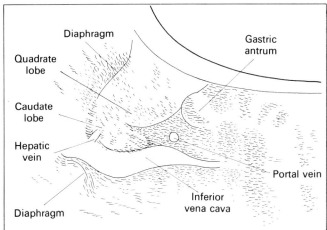

Fig. 7.8 Sagittal section in the plane of the vena cava. This forms the anatomical line of division between the left and right lobes of the liver. A hepatic vein is seen entering the vena cava immediately below the diaphragm. The portal vein is seen inferiorly, anterior to the vena cava.

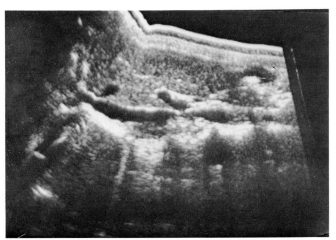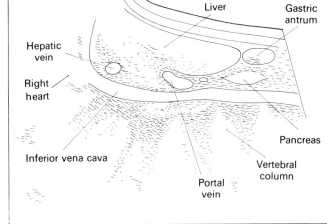

Fig. 7.9 Sagittal section in the plane of the vena cava. The liver is long, extending anterior to the portal vein which is dividing with right and left branches. An hepatic vein is seen below the diaphragm.

68 ULTRASONIC SECTIONAL ANATOMY

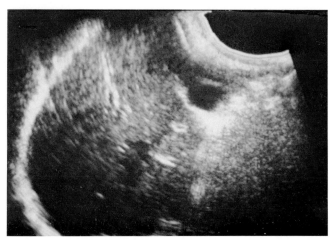 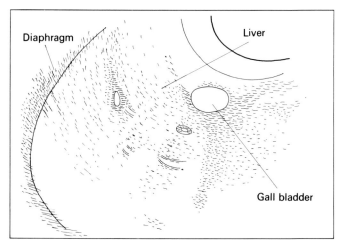

Fig. 7.10 Section in the right vertical plane. The liver is high, lying above the costal margin and textural detail is limited, even when recorded in full inspiration with low-frequency transducers and a correct swept gain slope.

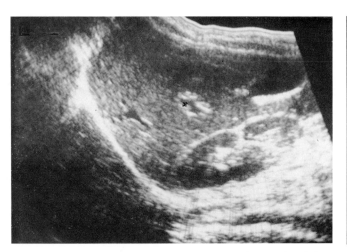 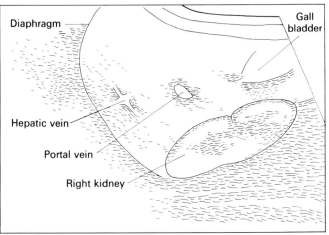

Fig. 7.11 Sagittal section in the right vertical plane with the liver extending below the costal margin. Textural detail is recorded and the right portal and right hepatic veins outlined.

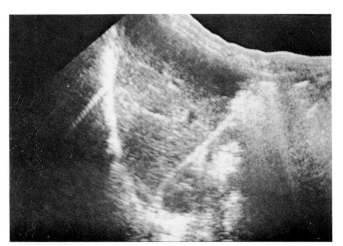 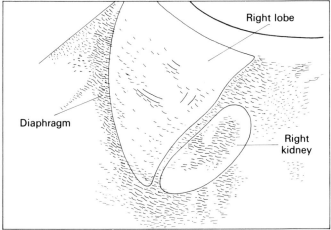

Fig. 7.12 Sagittal section lateral to the right vertical plane. The right lobe of the liver is short but textural detail is clear as the diaphragm is low and the liver descends below the costal margin in inspiration. The vessels are branches of the portal and hepatic veins.

UPPER ABDOMINAL VISCERA 69

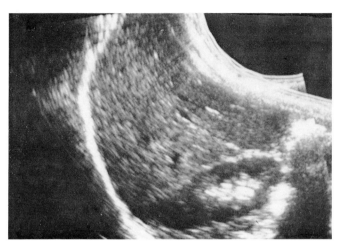 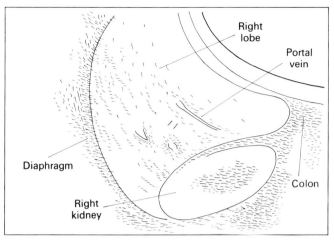

Fig. 7.13 Sagittal section lateral to the right vertical plane. The right lobe extends below the level of the lower pole of the right kidney. The portal vein seen in this section is a major unnamed branch supplying the right lobe which is frequently seen in lateral sagittal sections.

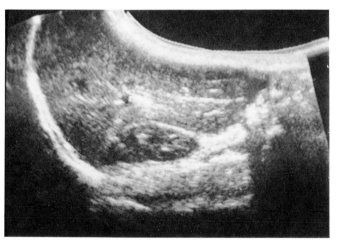 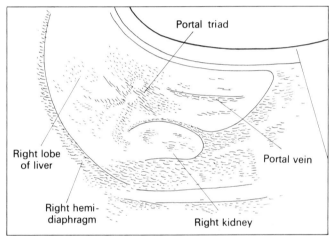

Fig. 7.14 Sagittal section with a long extension of the right lobe known as Reidel's lobe. The branch of the right portal vein is again outlined and the portal triad is seen centrally surrounded by Glisson's capsule.

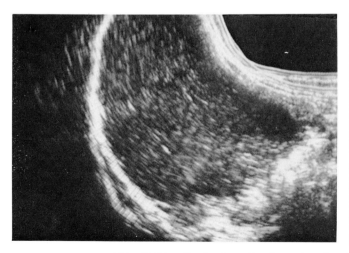 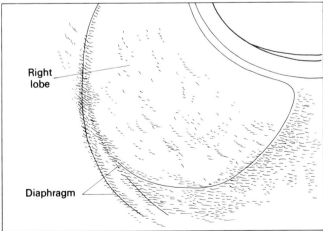

Fig. 7.15 Sagittal section lateral to the right kidney through the lateral aspect of the right lobe. Small vascular structures are defined. Posteriorly there is a double insertion of the diaphragm.

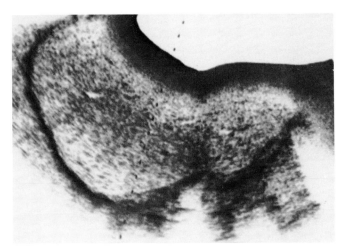 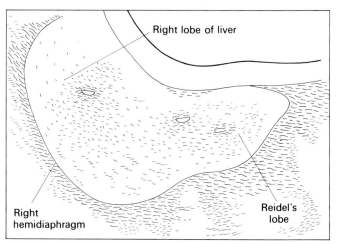

Fig. 7.16 Lateral sagittal section with a lobulated lower edge to the right lobe—Reidel's lobe. Veins are defined in the parenchyma.

Figures 7.17 to 7.25 are a series of transverse sections taken at different planes through the liver.

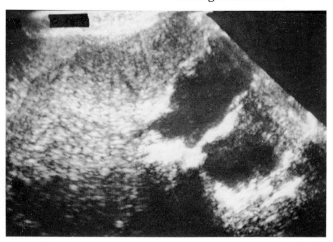 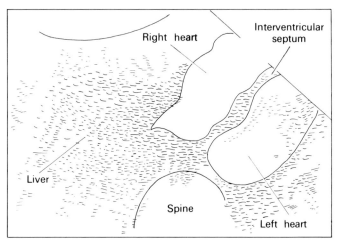

Fig. 7.17 High subxiphisternal section with cephalic angulation. On the right the liver is seen below the diaphragm. On the left side the section passes above the diaphragm through the heart.

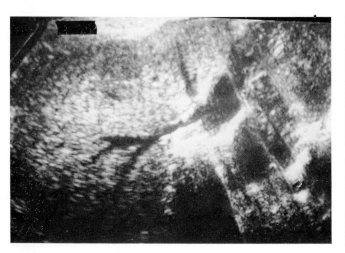 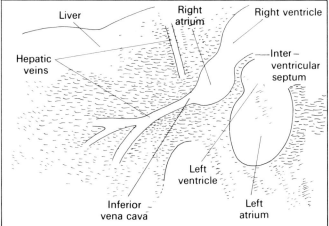

Fig. 7.18 Oblique subxiphisternal section passing below the right hemidiaphragm, through the central tendon and above the left dome of the diaphragm. The hepatic veins are seen in the liver joining the inferior vena cava just below the diaphragm; the vena cava then passes through the diaphragm into the right atrium.

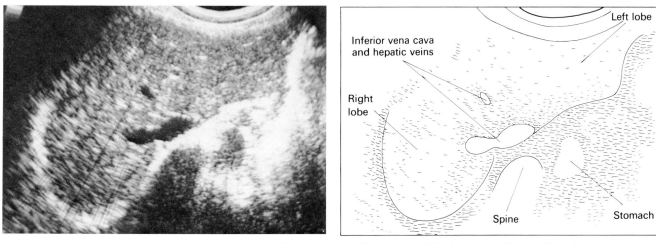

Fig. 7.19 High transverse section below the diaphragm. The hepatic veins and inferior vena cava are seen at this level.

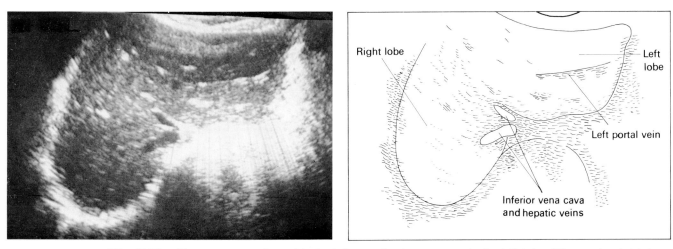

Fig. 7.20 Transverse section with the left portal vein now seen in the left lobe. The section is high as the hepatic veins are also seen entering the inferior vena cava.

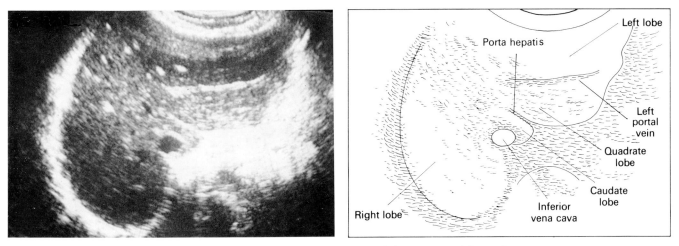

Fig. 7.21 Transverse section with the caudate lobe separated from the quadrate lobe by the lesser omentum in the porta hepatis.

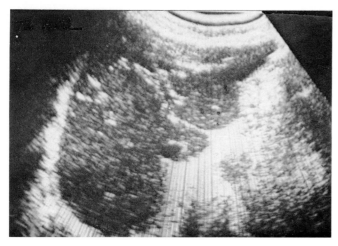 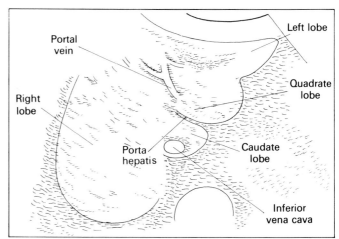

Fig. 7.22 Transverse section with unusual configuration of the posterior aspect of the left lobe. Note that the texture of the liver in this region remains homogeneous. Branches of the left portal vein are seen anteriorly.

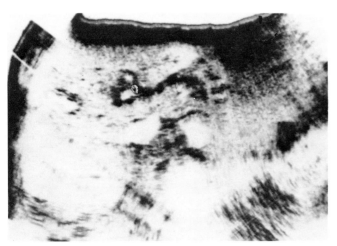 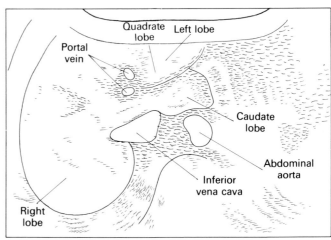

Fig. 7.23 Transverse section. The caudate lobe is of unusual configuration and size, extending anterior to the abdominal aorta. It is separated from the left lobe of the liver by a large fascial plane. Same subject as Figure 7.7.

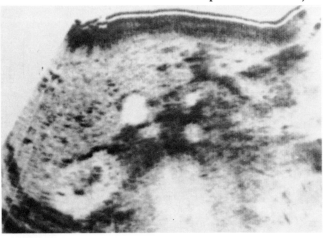 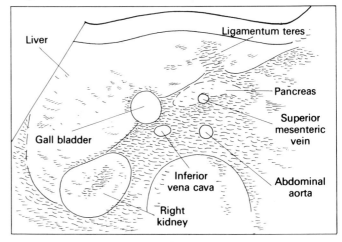

Fig. 7.24 Low transverse section below the porta hepatis. The fissure for the ligamentum teres is seen on the visceral surface of the left lobe. The gall bladder is to the right side of the midline, under the visceral surface, lying in the fossa for the gall bladder. The ligamentum teres divides the left lobe into medial and lateral segments, and the gall bladder lies in the plane dividing the anatomic right and left lobe.

UPPER ABDOMINAL VISCERA 73

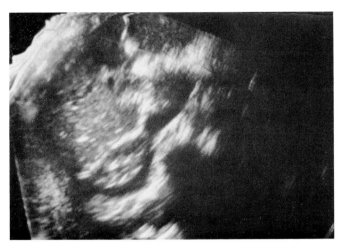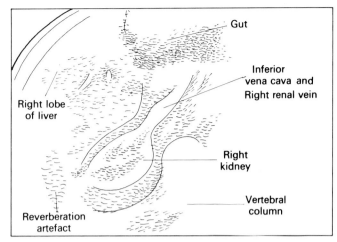

Fig. 7.25 Low transverse section below the left lobe with only the right lobe seen laterally. This section was taken through the right lower chest wall.

Figures 7.26 to 7.41 are a series of scans to demonstrate the hepatic veins and intrahepatic portal system.

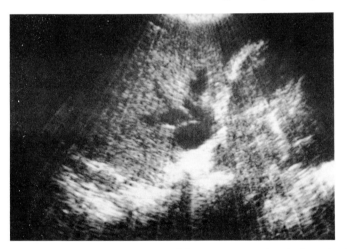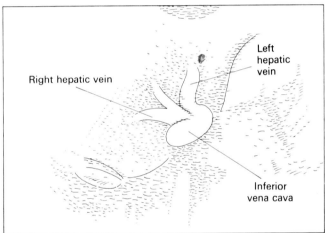

Fig. 7.26 Transverse section below the diaphragm with the major hepatic veins entering the inferior vena cava. The middle and left hepatic veins frequently unite before joining the vena cava.

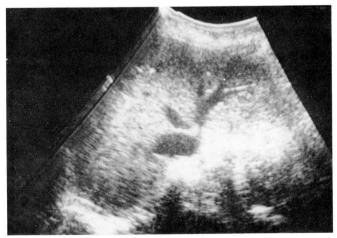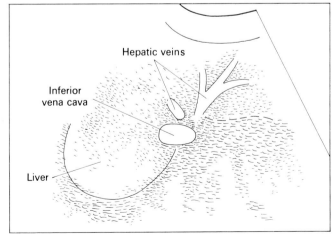

Fig. 7.27 Transverse section. The left and middle hepatic veins have joined before entering the inferior vena cava. There is an extra smaller hepatic vein anterior to the inferior vena cava. The right hepatic vein has not been sectioned.

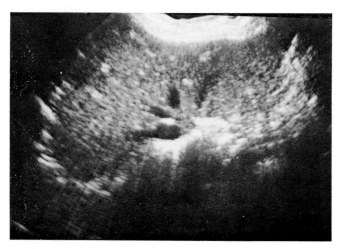 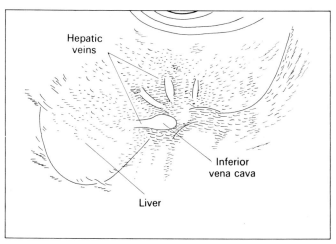

Fig. 7.28 Transverse section. There is a major right hepatic vein and several smaller veins. The number and configuration of the hepatic veins varies.

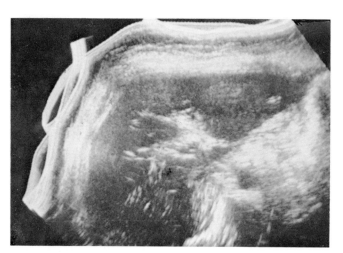 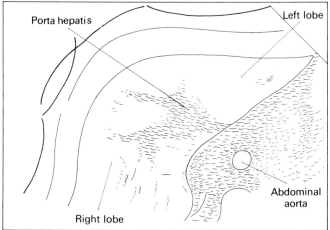

Fig. 7.29 Transverse section through the porta hepatis deep in the centre of the liver. The vessels and ducts are small at this level and only a mass of perivascular fibrous tissue is identified.

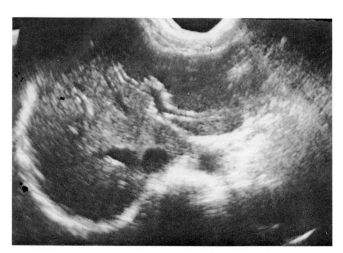 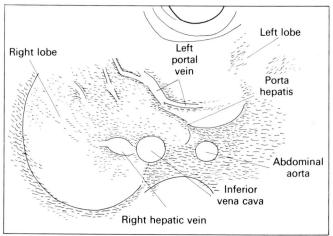

Fig. 7.30 Transverse section with cephalic angulation of plane. The left portal vein runs anteriorly in the porta hepatis sending branches to the medial and lateral segment of the left lobe. Portal veins are surrounded by connective tissue—an extension of Glisson's capsule. The hepatic veins do not have strongly echogenic walls as they are not surrounded by fibrous tissue.

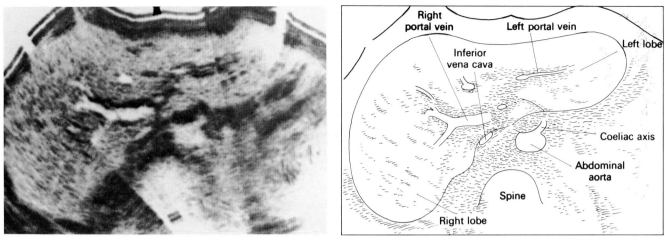

Fig. 7.31 Transverse section. The right portal vein takes a horizontal course into the right lobe before dividing into branches to supply the anatomic right lobe.

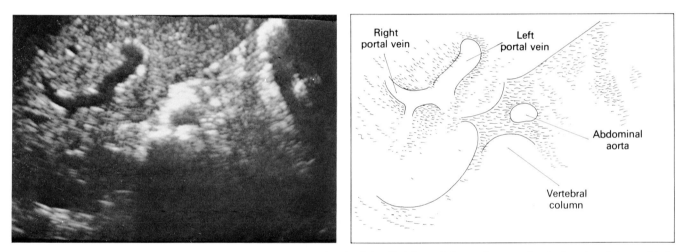

Fig. 7.32 Transverse section at the level of the division of the main portal vein into its right and left branches. The left branch curves forward and the right lies in a more horizontal plane.

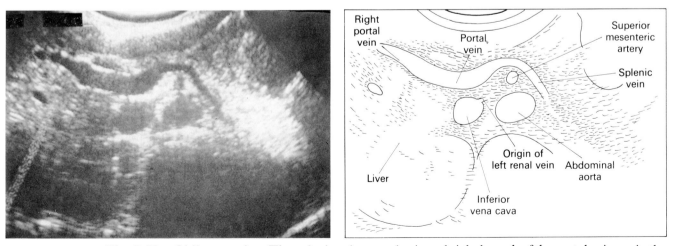

Fig. 7.33 Oblique section. The splenic vein, portal vein and right branch of the portal vein are in the same oblique plane.

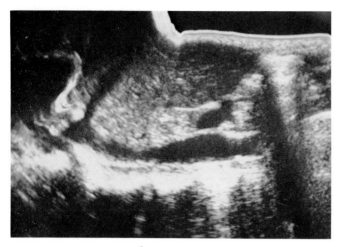 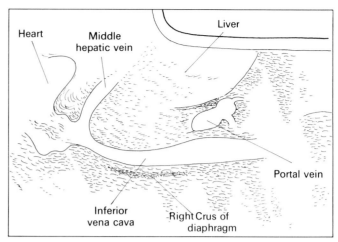

Fig. 7.34 Sagittal section in the plane of the inferior vena cava. The middle hepatic vein enters the vena cava anteriorly just below the central tendon of the diaphragm. The portal vein lies anterior to the vena cava—small tributaries are seen inferiorly.

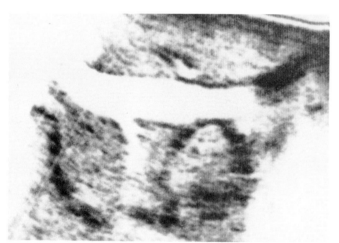 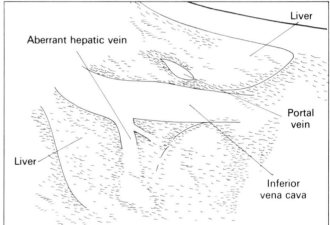

Fig. 7.35 Sagittal section with an aberrant hepatic vein entering the inferior vena cava posteriorly.

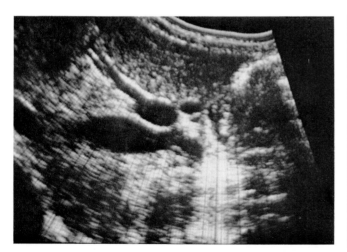 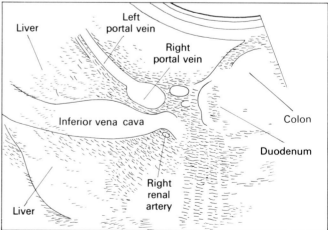

Fig. 7.36 Sagittal section at the division of the portal vein into its right and left branches. The exact position in which this division occurs varies. The left portal vein curves anteriorly into the left lobe, the right vein continues laterally in a horizontal plane. Note the position of the right renal artery behind the inferior vena cava.

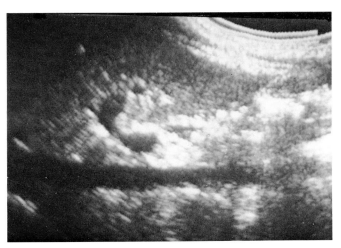 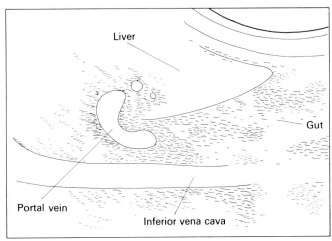

Fig. 7.37 Sagittal section. Typical inverted comma-shaped appearance of the portal vein at its division, with the left branch coursing anteriorly.

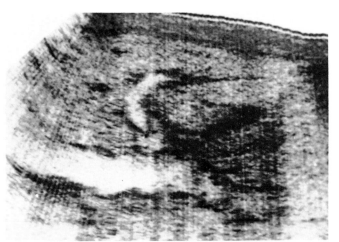 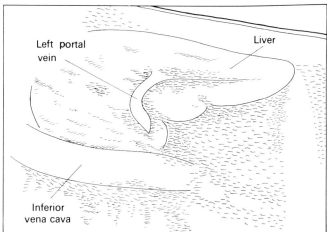

Fig. 7.38 Paramedian section. This branch of the left portal vein curves anteriorly and to the left to supply the left lobe of the liver. Its smaller branches are identified only by the surrounding perivascular fibrous tissue.

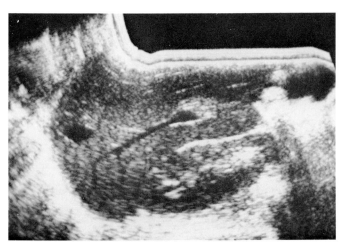 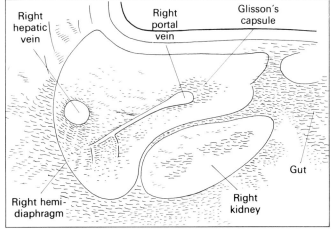

Fig. 7.39 Section in the right vertical plane. The right portal vein is dividing into smaller branches to supply the posterior area of the right lobe. The major right hepatic vein lies below the dome of the diaphragm.

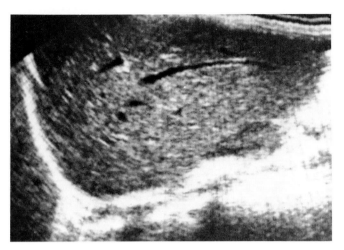 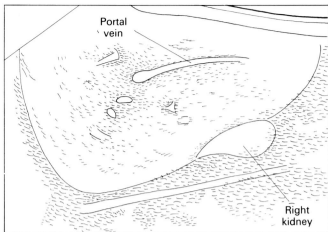

Fig. 7.40 Sagittal section lateral to the right median plane with a large branch of the portal vein supplying the inferior area of the right lobe.

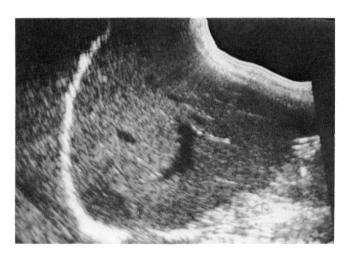 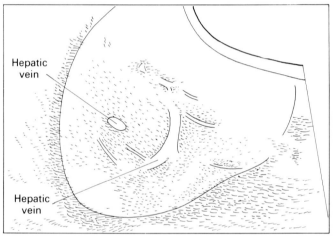

Fig. 7.41 Sagittal section. Large hepatic veins in the right lobe. These are identified by following their course medially to their major hepatic vein. Note that there is no condensation of perivascular connective tissue around the hepatic veins. The smaller tubular structures are branches of the right portal vein.

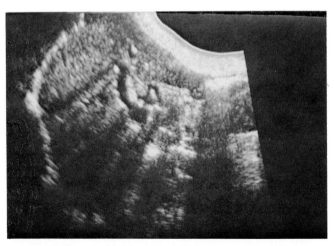 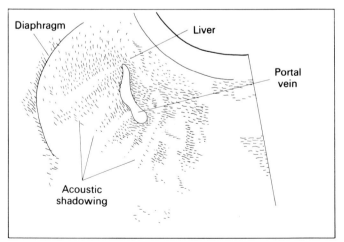

Fig. 7.42 Sagittal section of right portal vein with multiple areas of acoustic shadowing. Portal structures commonly cause shadowing which should not be confused with stones or gas in the biliary tree.

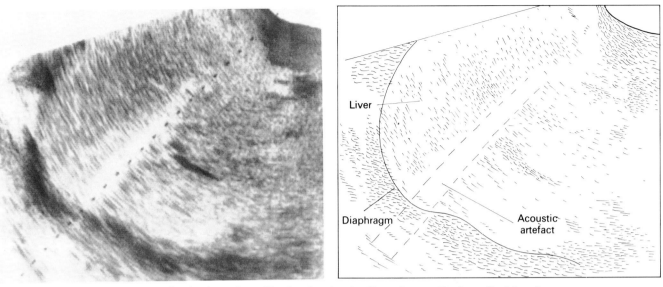

Fig. 7.43 Sagittal section. Shadowing in the liver from a Rodney Smith tube.

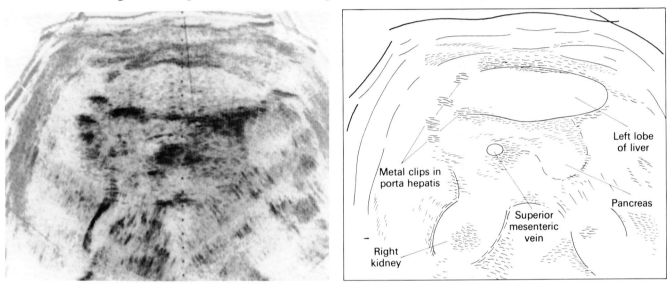

Fig. 7.44 Transverse section. Strongly reflective echoes produced by metal clips in the porta hepatis following cholecystectomy.

THE GALL BLADDER AND BILE DUCTS

The *gall bladder* is a thin pear-shaped organ covered by peritoneum and attached to the inferior surface of the quadrate and right lobes of the liver. Normally it is between 7–10 cm long and has a capacity of 30–60 ml. Anatomically it is divided into a *fundus*, a *body*, an *infundibulum called Hartmann's pouch* and a narrow *neck* leading into the *cystic duct*. The cystic duct is about 2–4 cm long and contains prominent mucosal spiral folds or valves of Heister. Anomalies of the gall bladder which are rare include absence, duplication, left-sided gall bladder and location of the gall bladder partially or completely in the liver. 'Floating gall bladder', in which the gall bladder is suspended by a mesentery, is the commonest anomaly.

The normal anatomical position of the gall bladder fundus is behind the ninth right costal cartilage at the junction of the costal margin with the right border of the rectus abdominis. On an ultrasonic section it is seen as an ovoid structure just to the right of the duodenum. It is usually related to the right portal vein which lies just superior to it. Since the gall bladder may have an oblique axis, it is best located by performing transverse sections and plotting out the degree of obliquity before performing longitudinal sections. Sometimes the gall bladder lies so close to the anterior abdominal wall that it is hard to separate from reverberations and a water-bath stand-off technique may be useful. Additional views, which can be used to bring out a gall bladder otherwise difficult to image, are a decubitus view with the left side down, and views taken in the erect position. These positions also help to differentiate viscid bile and calculi which can look very similar. Both can appear as a group of low-level echoes in the posterior aspect of the gall bladder. When the patient's position is changed, small gall-stones rapidly form a fresh fluid level whereas a level due to viscid bile takes many minutes to reform. The size and shape of the gall bladder are variable. There is no lower limit for size, though small gall bladders are often diseased. An upper limit for gall bladder size has variously been proposed as 200 ml; 5 cm in an antero-posterior direction and over 12 cm in length. None of these criteria is absolutely reliable. The gall bladder is frequently kinked or septated; such a finding may be normal but raises the question of hyperplastic cholecystosis. The normal gall bladder wall thickness is up to 3 mm though it may occasionally be greater, particularly if ascites is present.

The *bile duct canaliculi* which are located between the hepatic cells drain into progressively larger *intralobular and segmental bile ducts* which in turn drain into the *right and left hepatic ducts*. These unite in the porta hepatis to form the *common hepatic duct* which descends towards the duodenum. It is joined by the cystic duct to form the common bile duct. The right hepatic ducts are 1–2 cm long and join to form the common hepatic duct which is 2–4 cm long. The *common bile duct* is 8–15 cm long and 5–10 mm in outside diameter; it descends in the lesser omentum to the right of the hepatic artery and anterior to the portal vein. It passes behind the first part of the duodenum and through the pancreas to enter the descending duodenum, on its posteromedial aspect, at the ampulla of Vater where it is closely related to the pancreatic duct. Anatomical variations in the bile ducts, cystic artery and hepatic artery are common.

Only the larger components of the biliary tree can normally be visualised ultrasonically. The common duct is frequently seen and also the right and left hepatic ducts. The *common duct* (this is the term being adopted by sonographers for the common hepatic duct and common bile duct since the origin of the cystic duct can rarely be seen) can usually be found as it passes through the porta hepatis and across the anterior aspect of the foramen of Winslow in the hepatoduodenal ligament. At this point it lies anterior to and slightly to the right but in the same axis as the main portal vein. When it leaves the portal vein to enter the head of the pancreas it takes a posterior curve ending at the sphincter of Oddi. It has an echogenic outline similar to the portal vein and must be differentiated from the hepatic artery which lies just to the left and anterior to the portal vein. The hepatic artery gives off a large branch—the *gastroduodenal artery*—which takes a course through the head of the pancreas anterior to the common duct. The internal dimension of the bile duct is up to 6 mm diameter on an ultrasonic section, though following passage of a stone the common bile duct may not return to normal size and may remain dilated even though not obstructed.

UPPER ABDOMINAL VISCERA 81

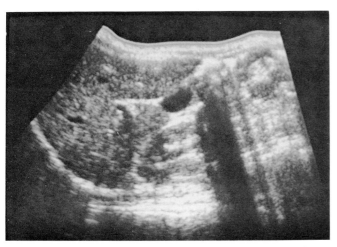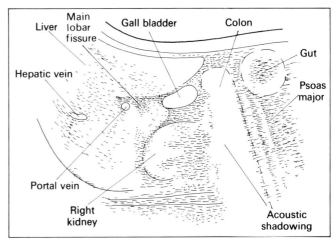

Fig. 7.45 Sagittal section through a normal relatively small gall bladder.

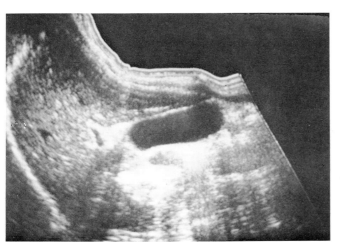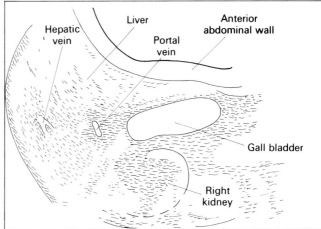

Fig. 7.46 Sagittal section. This gall bladder is 7 cm in length. Its anterior wall is defined clearly on this section and wall thickness can be measured.

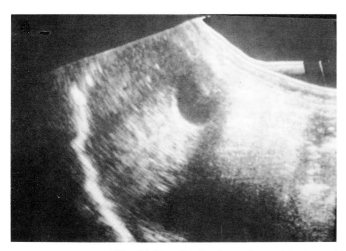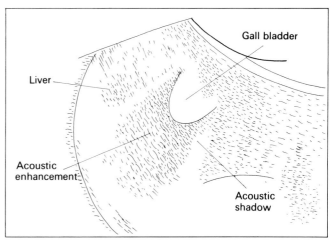

Fig. 7.47 Sagittal section with the gall bladder lying above the costal margin when the subject is supine. Note the strong distal central enhancement and peripheral shadowing which are the typical acoustic features of any cystic structure. Image detail is restricted. A gall bladder in this position is better examined with the subject erect or in the lateral decubitis position.

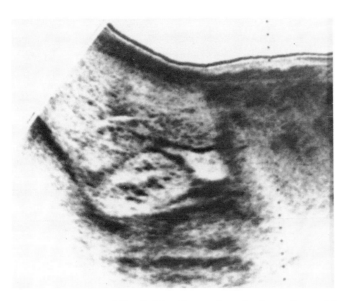 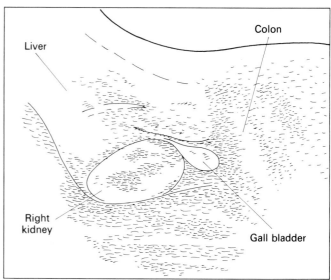

Fig. 7.48 Lateral sagittal section with the gall bladder lying below the kidney in a very lateral position.

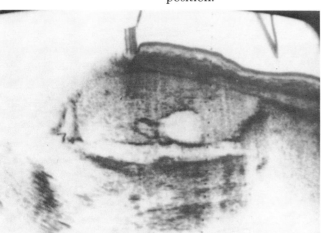 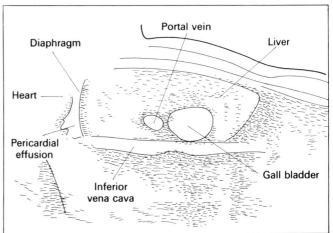

Fig. 7.49 Sagittal section in the plane of the vena cava. The gall bladder lies behind the liver, above its inferior edge and anterior to the inferior vena cava. It is partially intrahepatic.

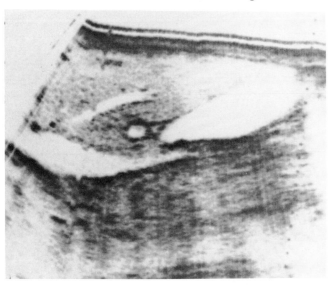 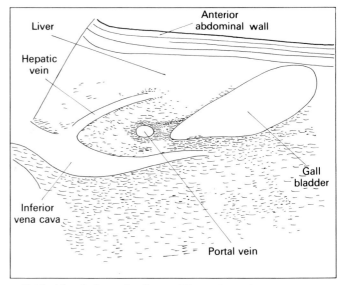

Fig. 7.50 Paramedian section with the gall bladder below the liver. There is a short mesentery suspending this gall bladder.

UPPER ABDOMINAL VISCERA 83

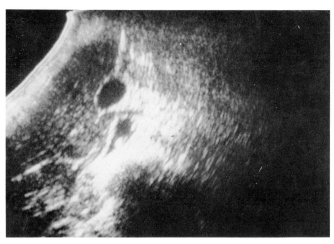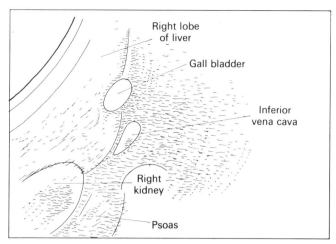

Fig. 7.51 Transverse section through the fundus of the gall bladder. This gall bladder lies below the costal margin, medial and anterior to the right kidney and anterior to the inferior vena cava. (Compare the position of this gall bladder with 7.52.)

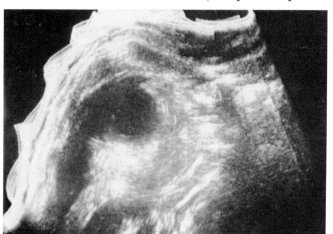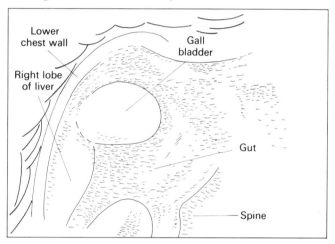

Fig. 7.52 Transverse section above the costal margin with the gall bladder in a high lateral position when the subject is supine. A gall bladder in this position is better examined with the subject erect or lying in the lateral decubitus position.

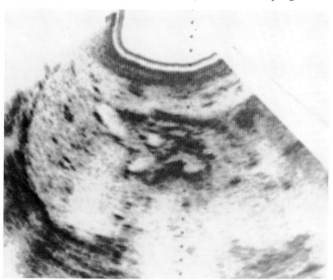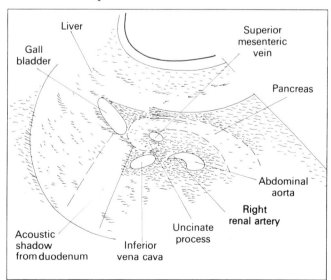

Fig. 7.53 Transverse section showing the normal anatomic relationship of the gall bladder to other structures in the right upper quadrant. The duodenum in this subject is identified by air in the first part producing an acoustic shadow. This may simulate stones in the gall bladder. The posterior wall of the stomach lies anterior to the pancreas and may be mistaken for the pancreatic duct. (See Fig. 10.17B.)

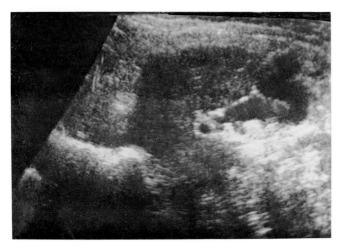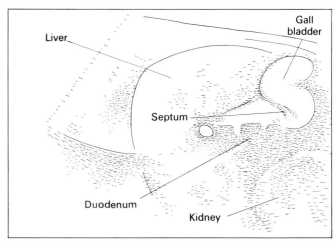

Fig. 7.54 Sagittal section with a kinked gall bladder and a partial septum in the neck. This partial septum is known as the junctional fold. The duodenum lies behind the gall bladder anterior to the right kidney. It is recognised by its irregular echogenic anterior border.

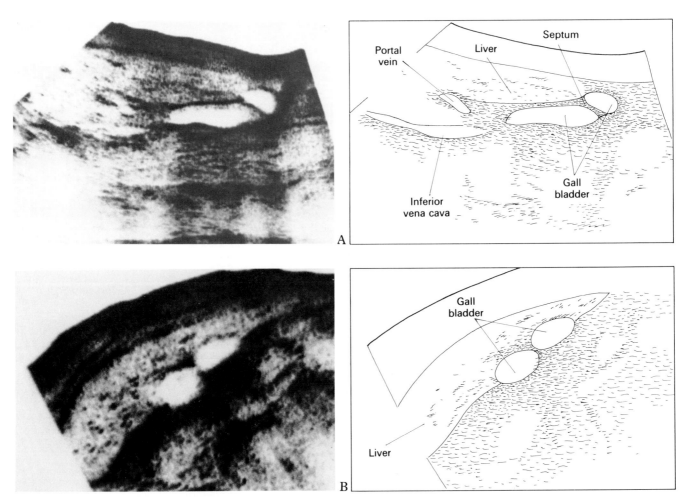

Fig. 7.55 Partial septum in the fundus—'Phrygian cap' deformity.
A. Sagittal section B. Transverse section.

UPPER ABDOMINAL VISCERA 85

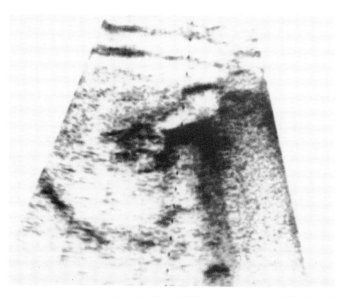 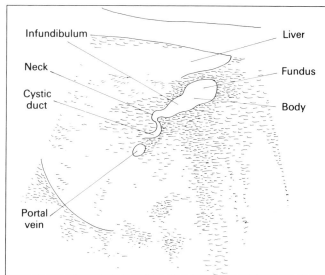

Fig. 7.56 Oblique section with detail of gall bladder and cystic duct.

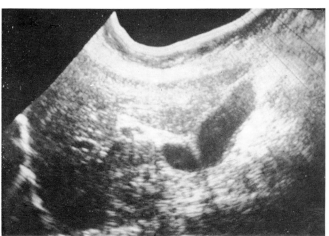 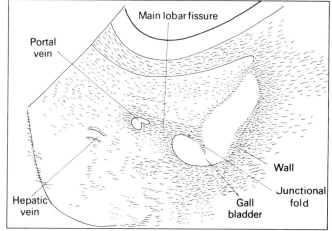

Fig. 7.57 Sagittal section. The walls of this gall bladder are well defined—they are approximately 3 mm thick which is within normal defined limits.

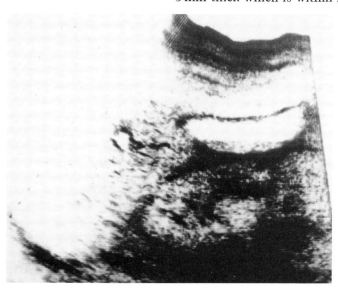 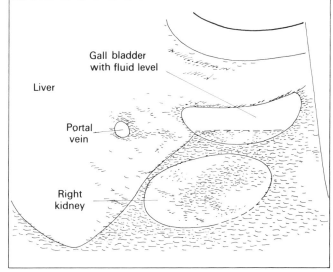

Fig. 7.58 Supine sagittal section. The gall bladder contains a fluid level formed posteriorly from sludge or viscid bile. This subject had no symptoms relating to the biliary tract and the finding was incidental.

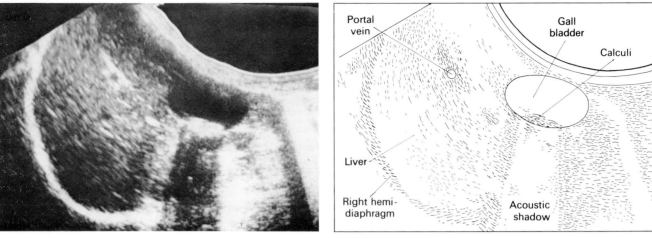

Fig. 7.59 A gall bladder containing several calculi. This subject also had no symptoms relating to the biliary tract.

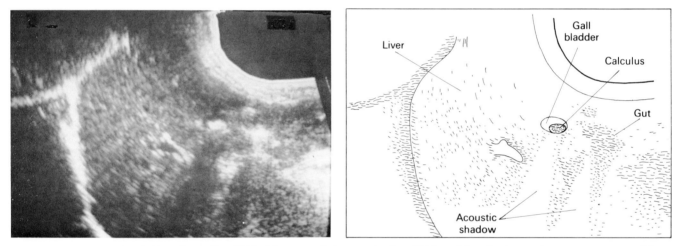

Fig. 7.60 A small gall bladder containing a calculus. This subject had no relevant symptoms. Note the similarity between the strong reflection and acoustic shadow produced by the calculus and gas in the intestine.

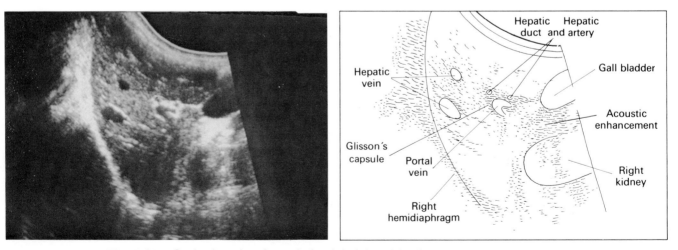

Fig. 7.61 Sagittal section through the right lobe of the liver with detail of the portal triad. The portal vein lies behind the right hepatic duct and right hepatic artery. They are surrounded by the highly echogenic Glisson's capsule. This relationship is maintained in the free edge of the lesser omentum, the hepatoduodenal ligament.

UPPER ABDOMINAL VISCERA 87

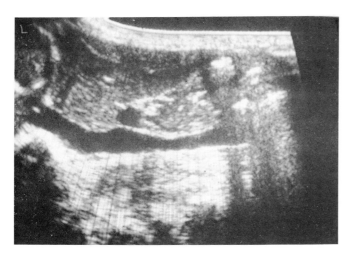 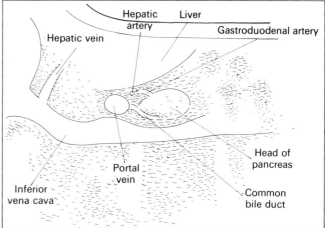

Fig. 7.62 Sagittal section. The common bile duct is anterior to the vena cava below the portal vein curving behind the head of the pancreas. The hepatic artery and gastroduodenal artery are more anterior.

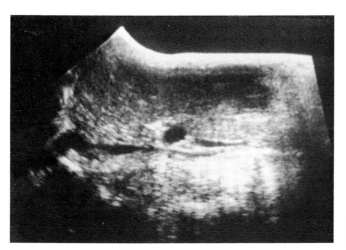 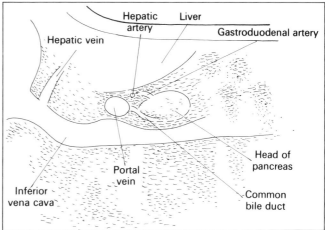

Fig. 7.63 Sagittal section. The bile duct is behind the head of the pancreas anterior to the vena cava. The gastroduodenal artery runs almost parallel to the bile duct in this section but anterior to the pancreas. The portal vein and hepatic artery are above the pancreatic head behind the left lobe of the liver.

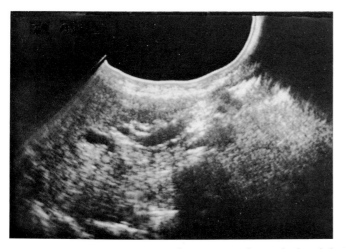 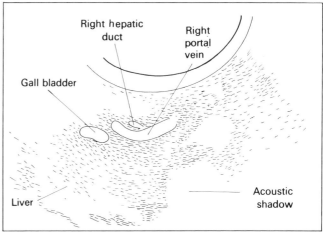

Fig. 7.64 Sector scan through the right lobe of the liver in the transverse plane. The right hepatic duct is anterior and parallel to the right portal vein. The vein is usually easily identified in the liver and the duct can then be located.

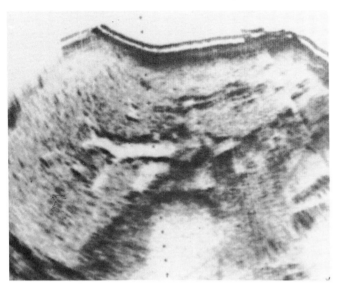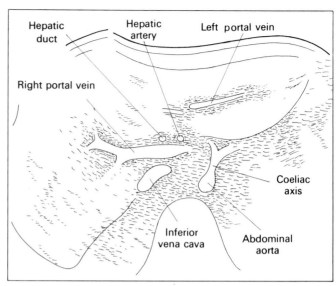

Fig. 7.65 High transverse section at the level of the coeliac axis and the entrance of the porta hepatis. The common hepatic duct and hepatic artery lie anterior to the portal vein.

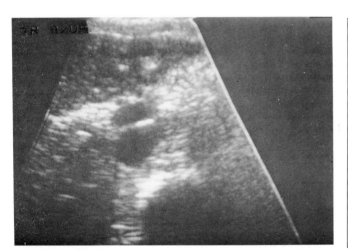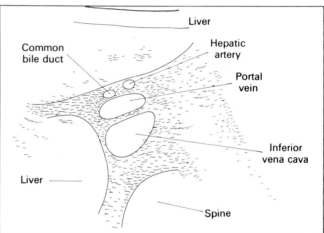

Fig. 7.66 Transverse section through the free edge of the lesser omentum. The common duct, hepatic artery and portal vein are separated from the inferior vena cava by the foramen of Winslow. (See 10.44.)

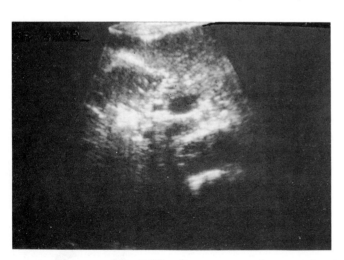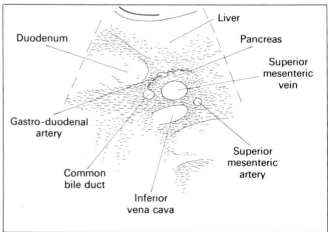

Fig. 7.67 Sector scan at the upper border of the neck of the pancreas. The gastroduodenal artery and bile duct are diverging to run respectively anterior to and posterior to the pancreas.

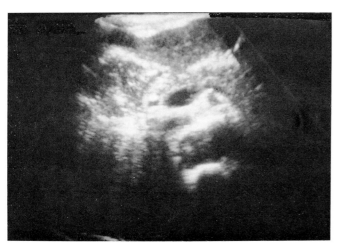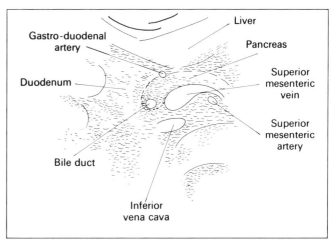

Fig. 7.68 Transverse section with the gastroduodenal artery and bile duct separated by the pancreatic head. The duodenum lies to the right of the pancreas and is distended with fluid and food debris.

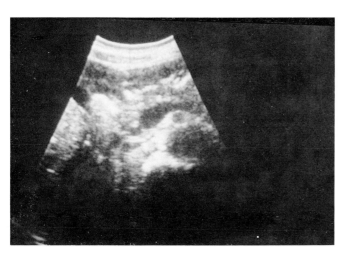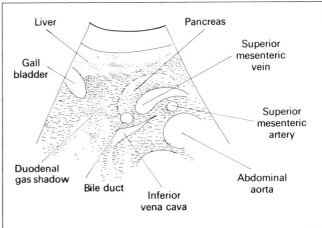

Fig. 7.69 Transverse section. The lower end of the common bile duct lies in the pancreatic head adjacent to the duodenum and anterior to the inferior vena cava. This section is immediately above the choledochoduodenal junction.

THE PANCREAS

The *pancreas* lies transversely in the upper abdomen, extending from the duodenal loop on the right to the hilus of the spleen on the left. It lies usually in a retrogastric position and crosses the vertebral column just below the coeliac axis. The axis of the pancreas is variable; it may be oblique, transverse, horseshoe or doglegged in shape. It is divided into a *head* with an *uncinate process*, a *neck*, *body* and *tail*. It passes anterior to the vena cava, the aorta, the superior mesenteric vein and artery and lies slightly anterior to or just below the splenic artery and vein. The head which lies just to the left of the descending duodenum is grooved posteriorly by the common bile duct.

There are two *pancreatic ducts*. The main duct begins in the tail and runs through the body slightly superior to the centre, receiving small tributaries. At the head it bends inferiorly, communicating with the accessory duct, and usually joins the bile duct as it pierces the duodenal wall. There are however variations in the size and anatomy of the pancreatic ducts and in their relationship to the common bile duct.

In the young adult normal pancreatic tissue is slightly more echogenic than the normal liver. Several landmarks make it possible to identify the location of the pancreas even if pancreatic tissue cannot be seen separate from adjacent structures. These landmarks include the inferior vena cava, the splenic vein, the superior mesenteric artery and vein and the common bile duct and gastroduodenal artery.

As the axis of the pancreas varies and since the pancreas can usually be found anterior to the inferior vena cava and the aorta, longitudinal sections are first taken at the level of the inferior vena cava and aorta and the pancreas marked. Transverse sections are then performed along the plotted axis. The main pancreatic duct can often be made out as a central line within the pancreas; it must be distinguished from the splenic vein posterior to the pancreas and the gastric antrum anterior to the pancreas. Normal upper limits of size have been proposed for the head and tail (2.5–3.0 cm) and body (1.5–2.0 cm) (Weill 1977, de Graaf et al 1978) but the pancreas may well be smaller and yet enlarged above its normal size. The normal pancreas shrinks with age and becomes more echogenic in normal subjects over the age of about fifty. It is easy to see in young adults but becomes increasingly difficult to see, because it is smaller, in older subjects. It is not uncommon for gas to hinder visualisation of the pancreas, particularly the tail. Visualisation can be improved by giving water and a peristalsis inhibitor such as fat, glucagon or probanthine. Sometimes it is possible to displace the gas with the examining transducer and both caudad or cephalad angulation of the scanning plane should be utilised. The erect or semi-erect position has also been found helpful for finding the pancreas if gas is present.

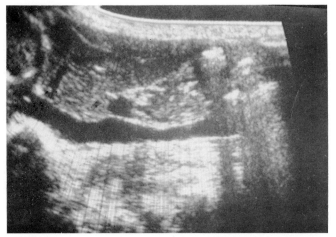 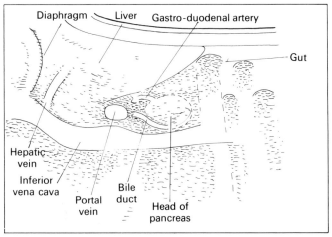

Fig. 7.70 Sagittal section with the pancreatic head lying anterior to the inferior vena cava and below the portal vein. The common bile duct grooves the posterior surface of the head and the gastroduodenal artery passes across its anterior surface.

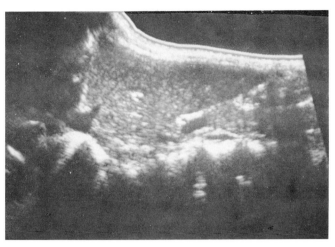 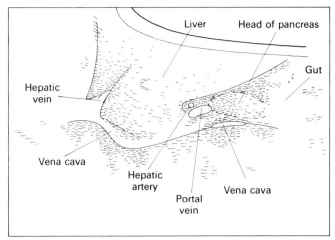

Fig. 7.71 Sagittal section through the medial aspect of the head of the pancreas. The texture of the pancreas is clearly imaged.

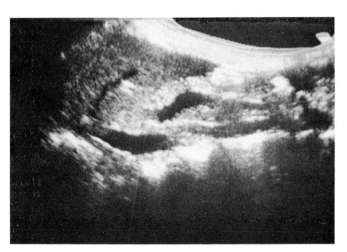 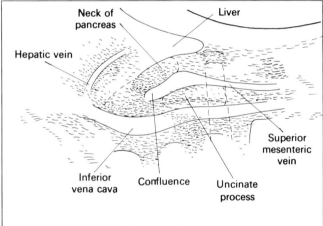

Fig. 7.72 Sagittal section. The superior mesenteric vein passes behind the neck of the pancreas to join the splenic vein. The portal vein is formed by the confluence of the splenic and superior mesenteric vein behind the neck of the pancreas. The uncinate process of the head extends medially behind the superior mesenteric vein.

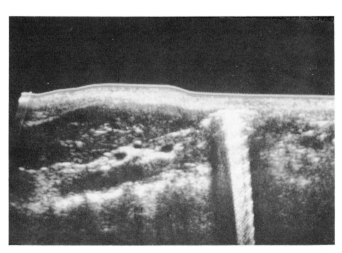 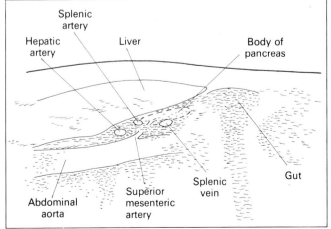

Fig. 7.73 Paramedian section. The body of the pancreas lies behind the left lobe of the liver anterior to the aorta and the superior mesenteric artery.

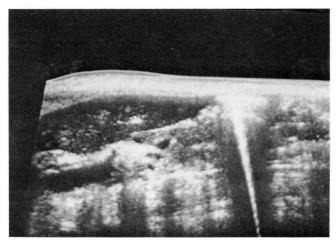 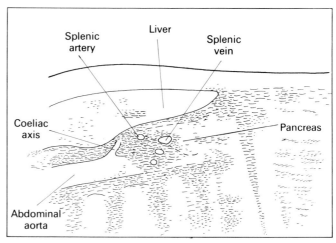

Fig. 7.74 Paramedian section. The coeliac axis and splenic vessels are above the tail of the pancreas which is obscured inferiorly by intestinal gas. The abdominal aorta is sectioned obliquely.

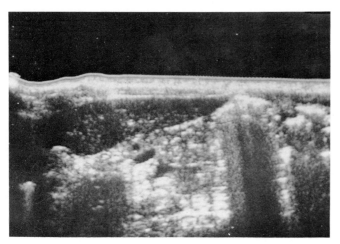 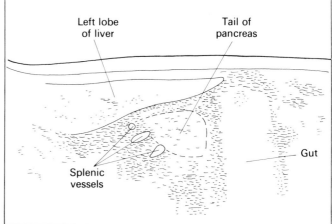

Fig. 7.75 Sagittal section to the left of the aorta. The tail of the pancreas is clearly imaged through the left lobe of the liver.

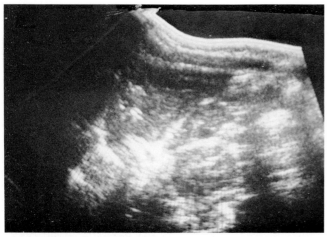 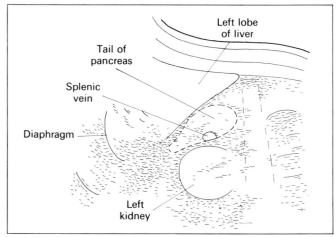

Fig. 7.76 Sagittal section lateral to the left vertical plane. The tail of the pancreas is lying anterior to the left kidney.

UPPER ABDOMINAL VISCERA 93

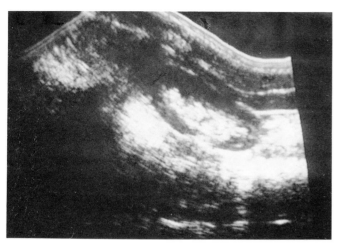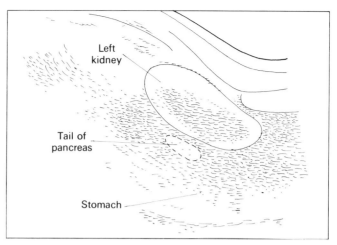

Fig. 7.77 Prone sagittal section. The tail of the pancreas is lying anterior to the kidney. Prone scans are used to examine the pancreatic tail when gas obscures detail with the subject in the supine position.

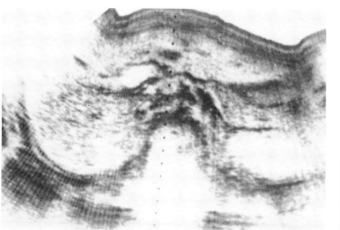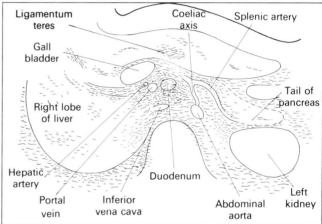

Fig. 7.78 High transverse section at the level of the coeliac axis. The tail of the pancreas is very clearly demonstrated. The axis of the pancreas is oblique and the body and head of the pancreas are inferior to this plane.

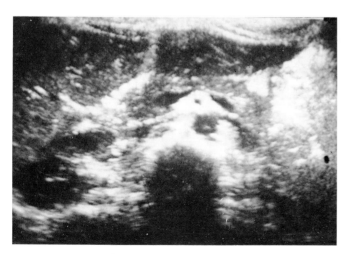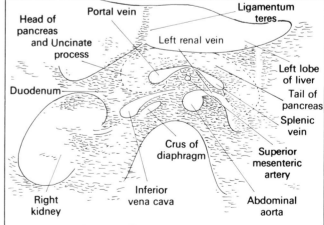

Fig. 7.79 Transverse section of the pancreas at the level of the splenic vein and the origin of the portal vein. The axis of this pancreas is transverse so that the head, neck, body and tail are outlined on the one section. The uncinate process of the head extends behind the superior mesenteric vein.

94 ULTRASONIC SECTIONAL ANATOMY

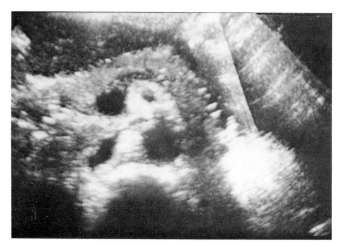 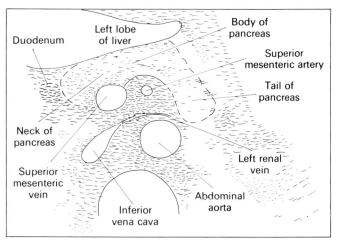

Fig. 7.80 Oblique transverse section below the splenic vein. The superior mesenteric artery and vein are behind the neck and body of the pancreas. The texture of the pancreas is particularly well demonstrated, and is more echogenic than the liver which is outlined anteriorly. The descending part of the duodenum is to the right of the pancreas but is not well defined.

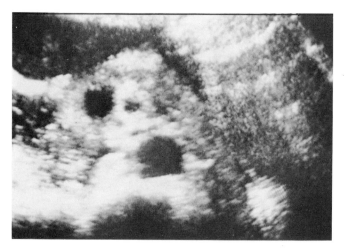 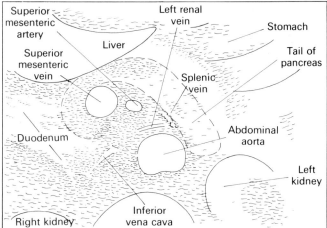

Fig. 7.81 Transverse section. The duodenum lies to the right of the pancreatic head. It contains fluid and food debris.

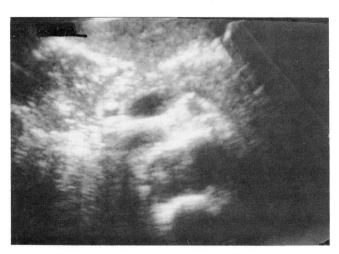 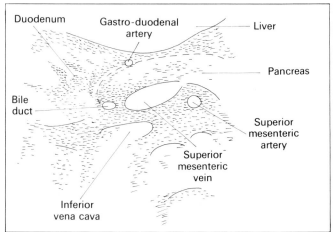

Fig. 7.82 Transverse section. The common bile duct lies behind the head of the pancreas. The gastroduodenal artery is anterior. These two structures identify the right border of the pancreatic head.

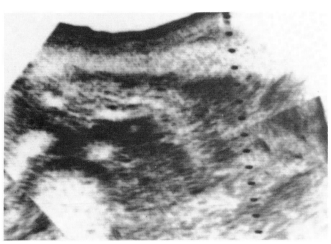 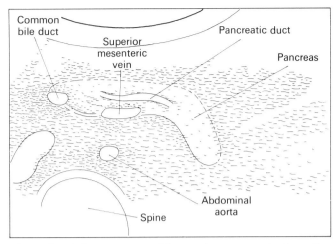

Fig. 7.83 Oblique transverse section in the axis of the pancreas. The main pancreatic duct lies centrally in the body of the pancreas. Below this level it bends inferiorly to pass through the neck and head of the pancreas into the duodenum at the ampulla of Vater. In this subject the duct is slightly dilated.

THE SPLEEN

The *spleen* lies in the left upper quadrant wedged between the diaphragm, stomach and left kidney. It is covered by peritoneum except at the hilus on the medial aspect of its concave gastric surface.

It is a difficult organ to examine ultrasonically unless enlarged, as it is partially obscured by lung and ribs. The best ultrasonic approach is to place the subject in the decubitus position, right side down, and perform sections in a longitudinal axis. Transverse sections are also possible but are of only limited value. Sections between the ribs in an oblique axis are advocated to assess texture, but the best textural detail is obtained using small sector scanners inserted in the intercostal spaces.

Within the otherwise homogeneous texture of the spleen central echoes may be seen near the medial surface representing the splenic hilum. The splenic artery and vein can be seen entering the spleen at this point.

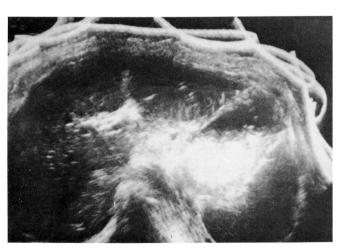 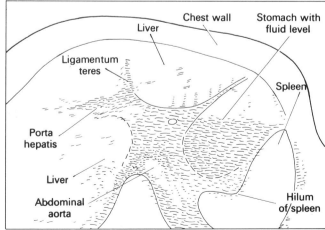

Fig. 7.84 Transverse supine section in the epigastrium with the spleen seen posteriorly behind the stomach. The stomach, which is filled with fluid and food debris, is acting as an acoustic window.

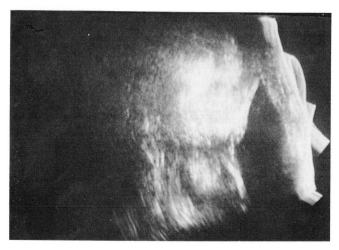 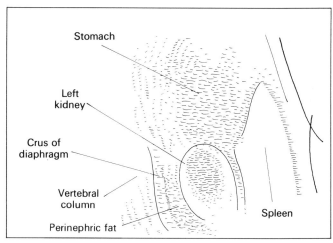

Fig. 7.85 Supine lateral section with the renal surface of the spleen lateral to the left kidney. This view can be used to assess the relationship of the tip of the spleen to the costal margin when assessing splenic size.

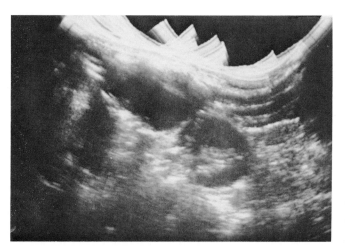 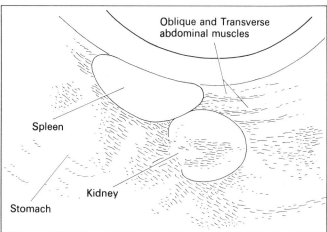

Fig. 7.86 Coronal section in the lateral decubitus position, right side down. This view is used to assess splenic size and texture. Textural detail is limited in the normal spleen as detail is obscured by overlying ribs and lung.

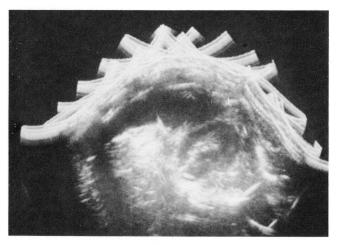 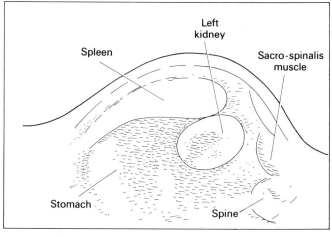

Fig. 7.87 Transverse section in the lateral decubitus position, right side down.

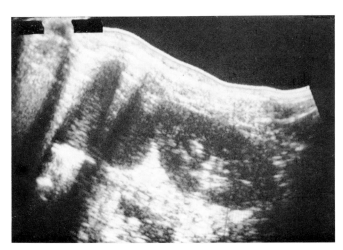 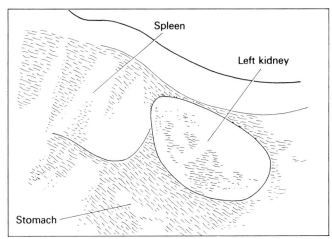

Fig. 7.88 Prone sagittal section. The spleen is partially obscured by the lower ribs, but between the reverberation artefacts there is a limited area of textural detail.

THE ADRENAL GLANDS

Each gland lies against the superior medial surface of its corresponding kidney.

The *right* gland is pyramidal in shape. It is wedged between the kidney, the diaphragm, the inferior vena cava and liver. The *left* gland is crescentic and is related to the stomach and pancreas anteriorly and to the diaphragm and kidney posteriorly. Each gland measures approximately 30–50 mm in height, 30 mm in breadth and from 4–6 mm in diameter.

The adrenals have an echogenicity greater than the kidney and liver and are only slightly less echogenic than the surrounding fibrous tissue. They are difficult to image and grey scale levels have to be optimally set to display the adrenals.

The *left adrenal* is located anteromedial to the upper pole of the left kidney adjacent to the aorta and spleen. It is best found by placing the subject in the right side down decubitus position, marking the site of the kidney transversely at several levels and then performing a longitudinal section along the axis of the kidney at an angle which shows the aorta medial to the kidney (Sample 1978).

The *right adrenal* gland is located superior and medial to the right kidney behind the inferior vena cava. A similar technique to that used on the left can be used also on the right side. The liver provides a transonic window which makes the right side easier to examine though the longitudinal shape of the adrenal is not easy to see. A laterally angled sagittal section through the inferior vena cava should show the adrenal, and transverse sections angling medially through the ribs to show the vena cava above the kidney should also demonstrate the right adrenal. The crus of the diaphragm lying posterior to the upper vena cava can easily be confused with the right adrenal.

Imaging of the normal adrenal glands is extremely difficult and time-consuming. Neither gland can be found all the time though the right is easier to find than the left. Ultrasonic assessment of the adrenal glands is frequently confined to examining the anatomical area occupied by the gland rather than demonstrating the gland itself.

98 ULTRASONIC SECTIONAL ANATOMY

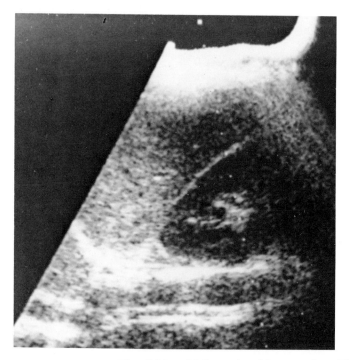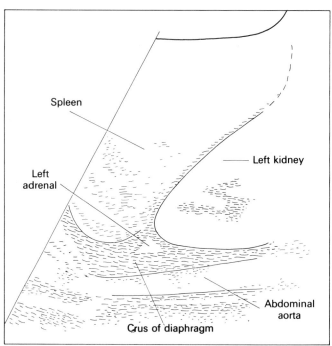

Fig. 7.89 Oblique decubitus section. The left adrenal gland is imaged between the spleen, kidney and aorta.

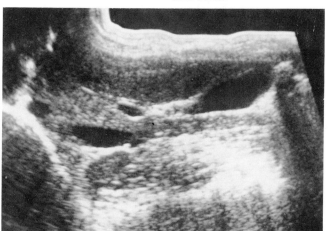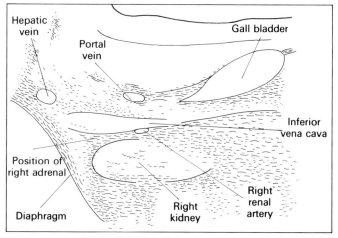

Fig. 7.90 Longitudinal section with the plane angled laterally through the vena cava and the upper pole of the right kidney. The right adrenal cannot be clearly defined but its anatomic position is indicated on the section.

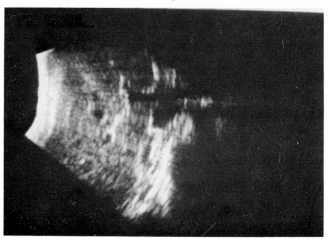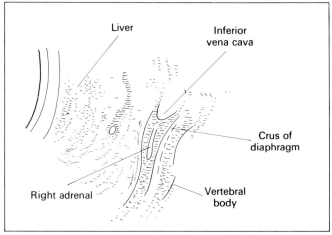

Fig. 7.91 Limited transverse section superior to upper pole of kidney. The right adrenal gland lies behind the vena cava, lateral to the right crus of the diaphragm and medial to the liver.

8 THE KIDNEYS

THE NORMOTOPIC AND ECTOPIC KIDNEY

The *developing kidney* migrates from the pelvis from the region of the fourth lumbar vertebra up to the level of the first lumbar vertebra. During its ascent it receives its blood supply and rotates so that the dorsal surface becomes the convex lateral border. Renal anomalies are relatively common—they are usually divided into anomalies of number, anomalies of position, anomalies of rotation, anomalies of volume and structure, and anomalies of the renal pelvis.

The *normotopic kidneys* lie in the upper part of the paravertebral gutters, posterior to the peritoneum, tilted against the structures on the sides of the last thoracic and first three lumbar vertebrae. The superior pole of the left kidney may lie as high as the 10th dorsal vertebra or as low as the second lumbar vertebra. Radiographic measurements indicate a length of 12.6 ± 0.8 cm and a width of 5.9 ± 0.4 cm in females and 13.2 ± 0.8 cm by 6.3 ± 0.5 cm respectively in males. Anatomically their dimensions are given as 10 cm in length, 5 cm in width and 2.5 cm in thickness. The radiographic measurements are approximately 1.3 times larger than those that are found ultrasonically which are not subject to magnification. The right kidney is slightly lower and smaller than the left kidney.

The kidney is divided into the peripherally situated cortex, containing the glomeruli and portions of the tubules, and the centrally located medulla composed primarily of portions of the tubules and collecting ducts.

The kidneys are ovoid in outline but the medial margin is deeply indented and concave at its middle, where a wide vertical cleft—the *hilus*—transmits the structures entering and leaving the kidney. The hilus leads into a space within the kidney, the *renal sinus*, which takes up a large part of the interior of the kidney. It contains the greater part of the pelvis, the calyces, fat, blood vessels, lymph vessels and nerve supply of the kidney.

Each kidney is enclosed in a dense fibrous capsule which is readily stripped from its surface. The capsule passes through the hilus to line the sinus and becomes continuous with the walls of the calyces. This capsule is surrounded by a fatty capsule—the perirenal fat—which fills the space inside the loosely fitting sheath of renal fascia, which encloses the kidney and adrenal gland. This renal fascia is connected laterally, superiorly and medially with the transversalis fascia, the diaphragmatic fascia and the fascia around the renal vessels. Inferiorly the walls of the sheath are loosely united and because of this the kidney may descend if the fatty capsule is absorbed.

Ultrasonic examination of the kidneys has traditionally been performed in the prone position. Transverse scans have been advocated because these identify the level and axis best followed by longitudinal sections, the axis being plotted from the transverse scans. Problems with the prone technique are particularly common on the left where ribs often interfere with adequate visualisation of the left upper pole. A supine longitudinal approach defines the right kidney well in most subjects and gives a much better image of the ultrasonic texture. The decubitus position has been found to be superior to the prone position in showing the upper pole of the left kidney and should always be used when information is restricted on supine and prone sections of either kidney.

The normal kidney is the least echogenic of the major upper abdominal organs. A group of dense echoes are seen in the centre of the kidney arising from the sinus fat, the renal arteries and veins and the walls of the pelvicalyceal system. Parenchymal detail can be seen with high quality images; the cortex is normally more echogenic than the medulla.

At the apex of the medulla it may be possible to see a central echo which represents the arcuate artery. It is also not uncommon to define the major vessels on supine scans; the renal artery and renal vein leave the kidney in an anterior-medial direction.

The degree of distension of the renal pelvis is variable. If large, the renal pelvis may be impossible to differentiate from mild hydronephrosis, unless the dilated pelvis can be followed into a dilated ureter. A semi-prone decubitus position is best for defining the details of these major pelvic structures.

Provided the axis has been plotted on transverse sections a satisfactory renal length can be obtained from the sonogram. It should be noted that this is normally shorter than the radiographic measurement because of the absence of magnification. Spuriously short lengths for kidney on the radiograph due to the forward tilt of the lower pole do not occur with sonography.

A number of normal variants may be recognised in the kidney including a bifid pelvicalyceal system, a horseshoe kidney, crossed renal ectopia, a pelvic kidney and an unusual axis tilt to the kidney.

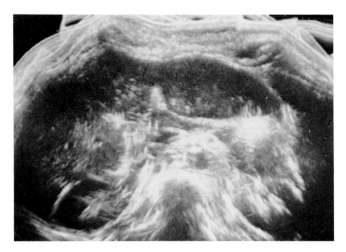 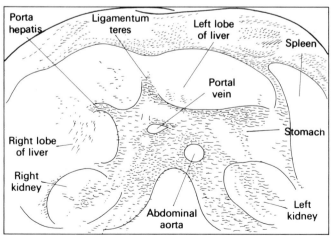

Fig. 8.1 High transverse supine section. Both kidneys are outlined, the right is clearly defined behind the right lobe of the liver and the left lies behind the stomach and medial to the spleen, which is slightly enlarged.

THE KIDNEYS 101

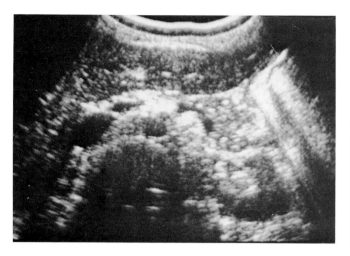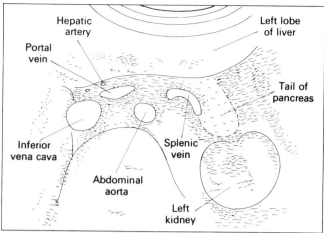

Fig. 8.2 Transverse supine section. The tail of the pancreas lies anterior to the left kidney.

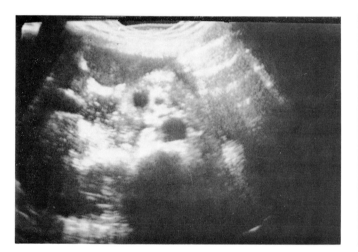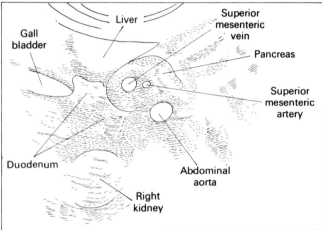

Fig. 8.3 Transverse supine section. The duodenum is lateral to the pancreatic head and anterior to the right kidney which is partially obscured by intestinal contents.

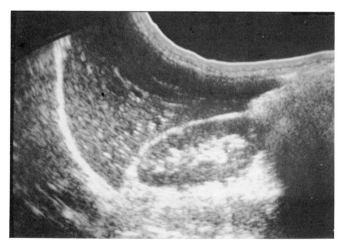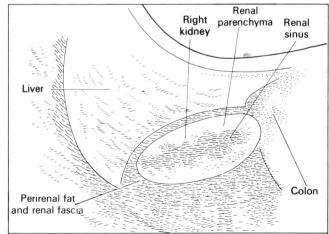

Fig. 8.4 Supine sagittal section. The right kidney is clearly outlined behind the right lobe of the liver.

102 ULTRASONIC SECTIONAL ANATOMY

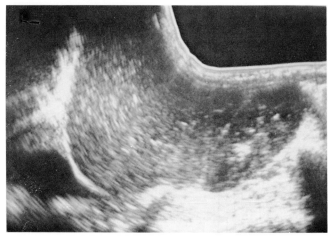 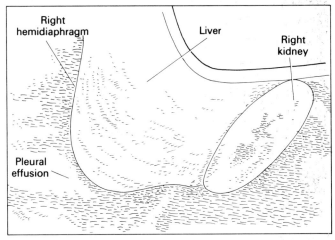

Fig. 8.5 Supine sagittal section with the right kidney below the right lobe of the liver. Note the difference of the angle in the longitudinal axis of the kidney in this section and the kidney in Fig. 8.4.

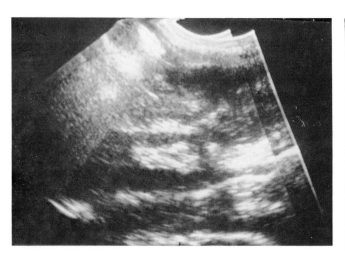 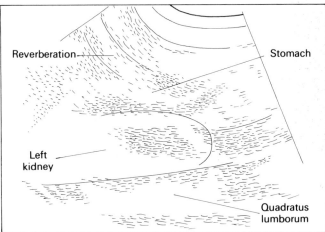

Fig. 8.6 Supine sagittal section lateral to the left vertebral plane. The left kidney is outlined behind the stomach and detail of the upper pole is obscured by gas in the fundus of the stomach causing marked reverberation artefacts.

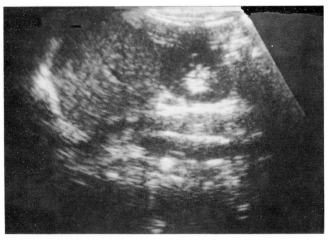 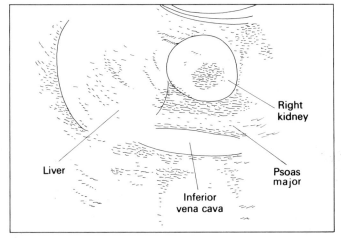

Fig. 8.7 Coronal section with subject in the lateral decubitus position (left side down). The plane of the scan is angled anteriorly passing through the right kidney, the psoas muscle and inferior vena cava.

THE KIDNEYS 103

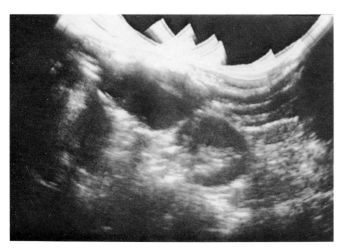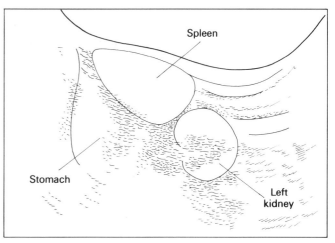

Fig. 8.8 Lateral decubitus (right side down) coronal section. The left kidney is outlined below the spleen. Textural detail is restricted by overlying ribs.

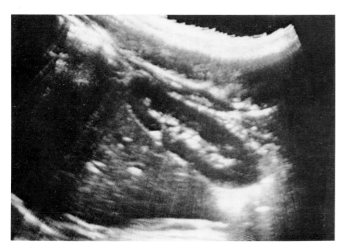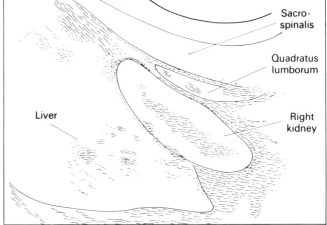

Fig. 8.9 Prone sagittal section through the longitudinal axis of the right kidney which measures 10.8 cm in this subject.

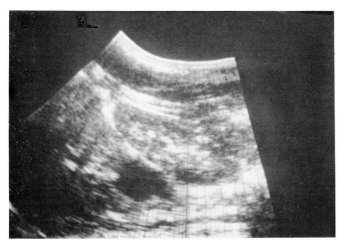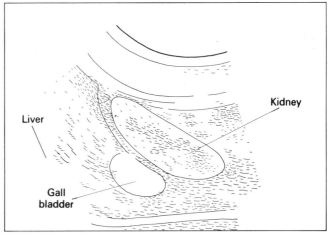

Fig. 8.10 Prone sagittal section. The gall bladder is seen ventral to the right kidney behind the right lobe of the liver.

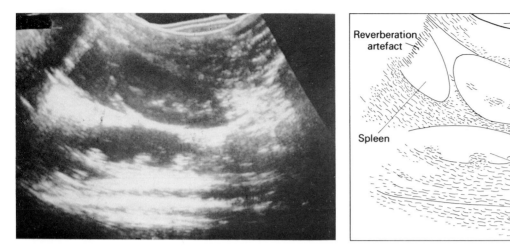

Fig. 8.11 Prone sagittal section. The fluid-filled stomach is ventral to the left kidney and the spleen.

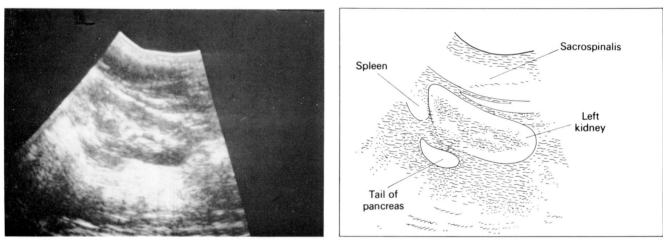

Fig. 8.12 Prone sagittal section. The tail of the pancreas lies anterior to the left kidney. The tip of the spleen is adjacent to the upper pole of the kidney.

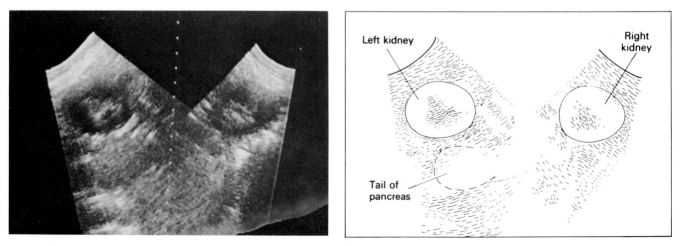

Fig. 8.13 Prone transverse section. Both kidneys are outlined. The pancreas is indistinctly outlined anterior to the left kidney.

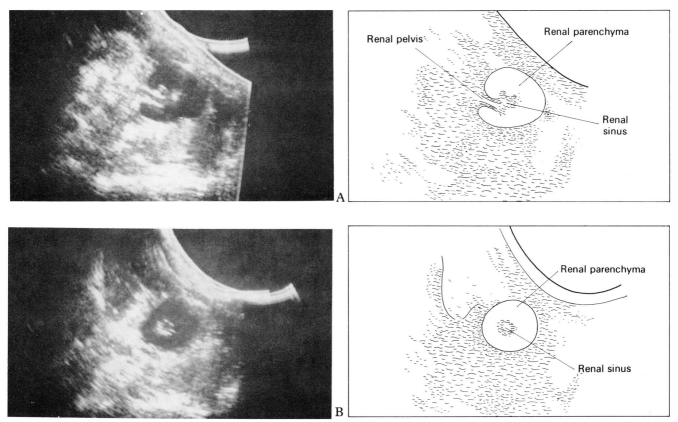

Fig. 8.14 Prone transverse sections of the right kidney. The renal sinus is medially placed at the level of the renal pelvis (A) and central in a section through the upper third or lower third of the kidney (B).

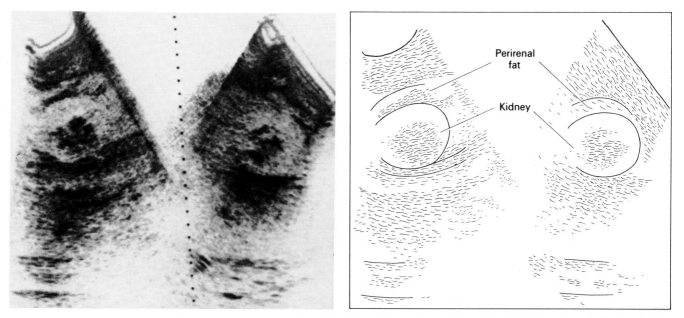

Fig. 8.15 Prone transverse section. In this subject the perirenal fat is defined posteriorly. It lies between the fibrous capsule of the kidney and the perirenal fascia.

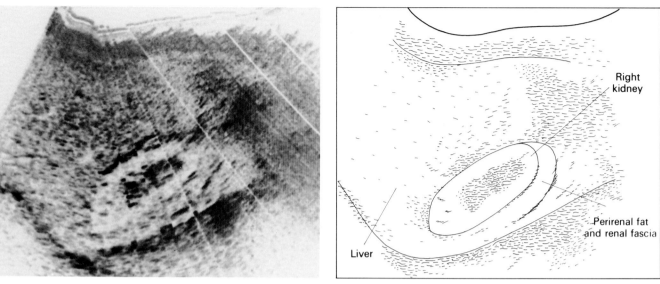

Fig. 8.16 Supine sagittal section lateral to the right vertical plane. Perirenal fat and fascia are defined behind the right kidney.

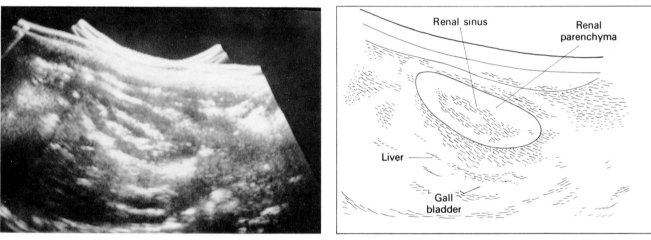

Fig. 8.17 Prone sagittal section. There is slight separation of the renal sinus echoes by a distended intrarenal pelvis. This degree of separation is normal and is frequently seen in subjects prepared for pelvic examination with oral hydration.

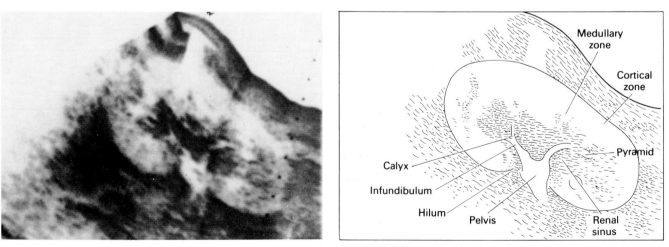

Fig. 8.18 Prone section with well demonstrated pelvicalyceal detail in the renal sinus. Parenchymal detail is not so well seen; the pyramids in the medullary zone are relatively echo-free, the cortex is more echogenic.

THE KIDNEYS 107

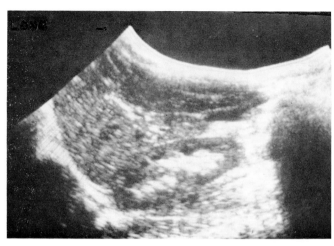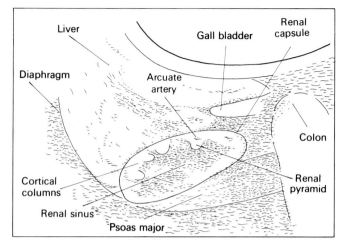

Fig. 8.19 Supine sagittal section of the right kidney. Parenchymal and renal sinus detail are demonstrated. The renal sinus is highly echogenic. The parenchyma is surrounding the sinus; the cortex and medulla are defined with the cortical columns between the pyramids. Arcuate vessels are identified.

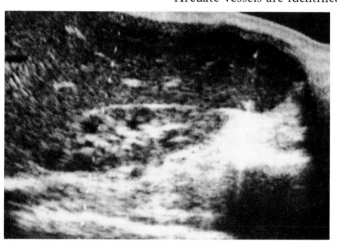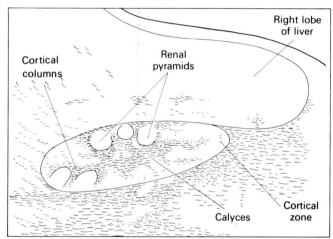

Fig. 8.20 Supine sagittal section of the right kidney. Parenchymal detail is well demonstrated and the renal sinus is thinner. The section is in a more lateral plane than Figure 8.19.

RENAL ANOMALIES

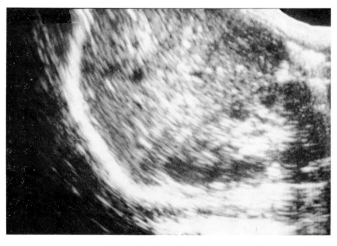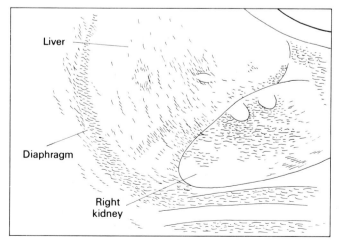

Fig. 8.21 Supine sagittal section. Right renal hyperplasia. Compare the size of this kidney with those in Figures 8.4, 8.5 and 8.19. Full investigation failed to identify a left kidney.

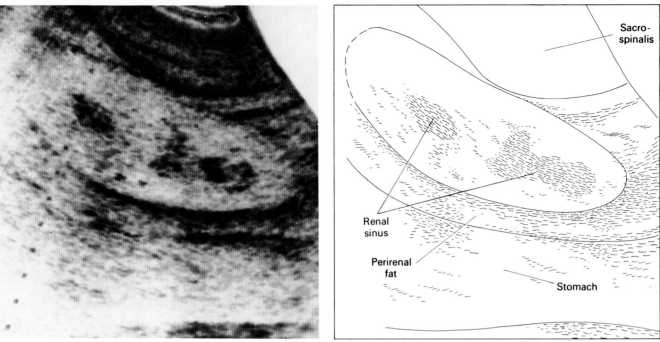

Fig. 8.22 Prone sagittal section. Duplex kidney. The kidney is long with a bifid sinus. On intravenous urography there was duplication of the renal pelvis—incomplete duplication of the kidney.

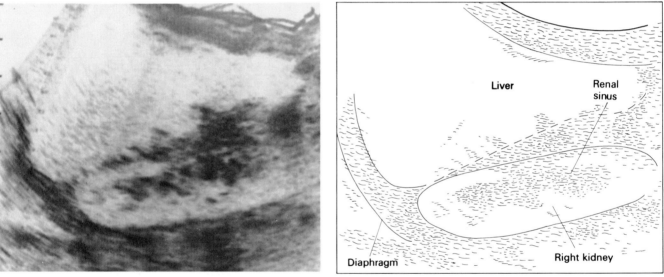

Fig. 8.23 Supine sagittal section. Malrotated right kidney. The renal sinus is anterior; the kidney has failed to fully rotate during its ascent from the pelvis.

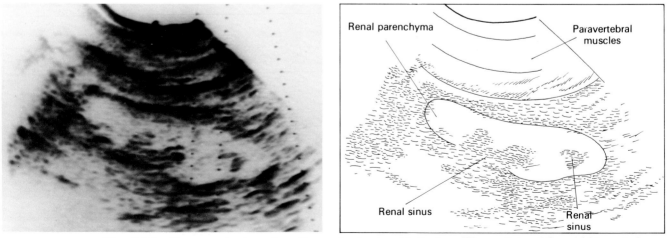

Fig. 8.24 Prone sagittal section. Crossed ectopia with fusion. The unilateral fused kidney is long, measuring between 14 cm and 15 cm in its long axis. There are two separate renal sinuses identified. The crossed ectopic kidney is fused to the lower pole of the normotopic kidney.

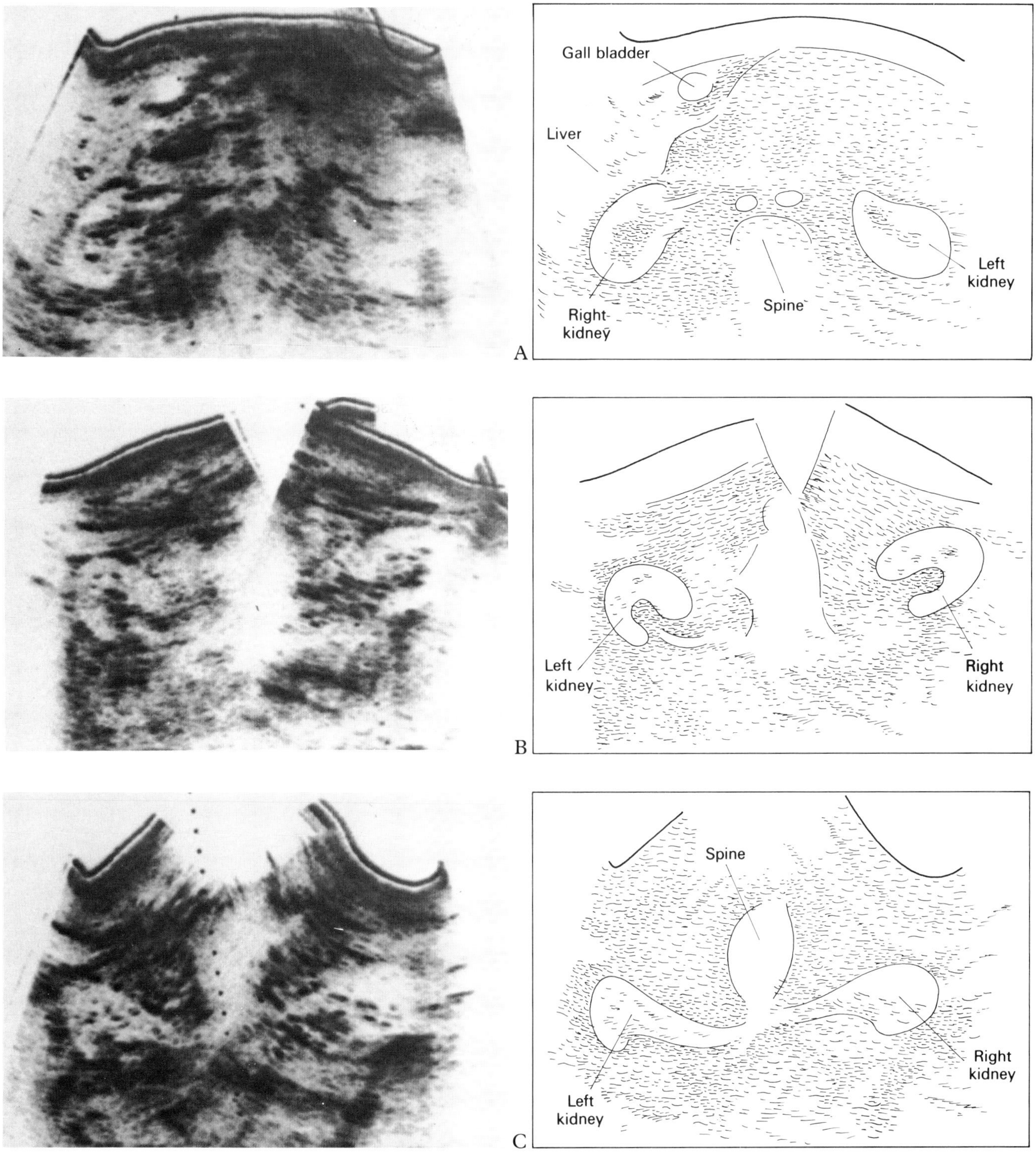

Fig. 8.25 Supine (A) and prone (B) and (C) transverse sections. Horseshoe kidney. This is the commonest form of renal fusion. There is anterior rotation of the renal sinus and a bar of tissue passing anterior to the vertebral column. It is better defined on the left side. This bar of tissue passes anterior to the aorta and behind the inferior vena cava, connecting the lower poles of the kidneys.

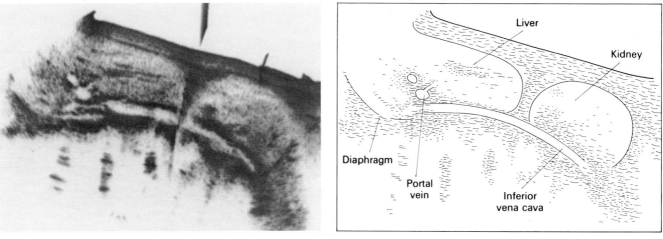

Fig. 8.26 Paramedian section. Pancake kidney in the lower abdomen anterior to the inferior vena cava. Pelvic kidneys, so long as they are not diseased, look like a normal kidney. Pancake kidneys have a similar location and pattern but the pelvicalyceal echoes may be less obvious.

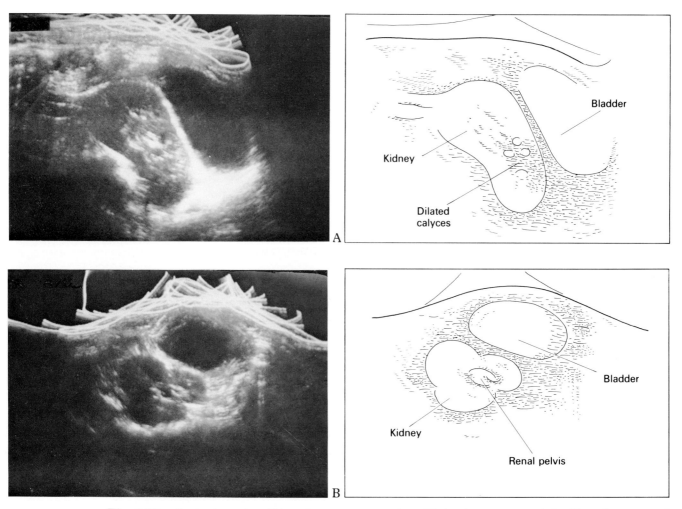

Fig. 8.27 Sagittal section (A) and transverse section (B) in the greater pelvis. Ectopic presacral kidney with dilated calcyces.

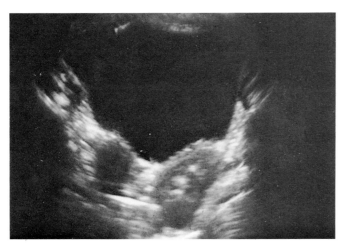 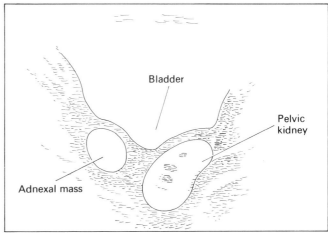

Fig. 8.28 Transverse section in the lesser pelvis. Left-sided pelvic kidney, with normal structure and normal function. Recognition of a pelvic kidney is particularly important when assessing the pelvis for possible pathology. Note the small right-sided associated adnexal mass.

THE TRANSPLANTED KIDNEY

The *donor kidney* is placed retroperitoneally in the iliac fossa. The left kidney is preferred because of its longer renal vein. A left kidney is usually implanted in the right side and vice versa, though either kidney may be implanted in either iliac fossa if the iliac vein is mobilised by division of the hypogastric vein. The kidney is rotated 180° so that the posterior surface lies anteriorly. Anastomosis is carried out between the renal artery and hypogastric artery (end to end) and between the renal vein and external iliac vein (end to side). Ureteroneocystostomy is usually accomplished through a surgically constructed submucosal tunnel. If the donor ureter cannot be utilised ureteropyeloplasty or uretero-ureterostomy are alternative procedures.

Adult kidneys may readily be implanted into children either retroperitoneally or transperitoneally with end to side anastomosis with the common iliac vessels or the aorta and vena cava.

Renal transplants are particularly well suited for ultrasonic study as they are superficially situated with no interposed bowel, gas or bony structures. Transducers of 3.5 MHz or 5.0 MHz are routinely employed, producing high quality images.

Patients may be scanned as early as 24 hours after transplantation. No special precautions are required with regard to transducer sterility, but a thin layer of polythene interposed between the transducer and the skin, coupled to the skin with sterile oil or gel, will reduce the possibility of cross-infection. The polythene also reduces discomfort, if scanning over stiches or around drainage tubes is necessary. If possible it is preferable for fluid to be present in the bladder, but in the majority of patients examined after recent transplantation the bladder is empty. If a catheter is already in the bladder sterile fluid may be cautiously introduced. With static B scanners multiple transverse sections are used to determine the long axis of the kidney which is very varied; using dynamic scanners the examination is modified and rapid screening is possible.

In the transplanted kidney the normal renal configuration is preserved. The renal outline is smooth, the sinus echoes compact and parenchymal detail clearly imaged. The renal vessels may be imaged but detail is limited.

112 ULTRASONIC SECTIONAL ANATOMY

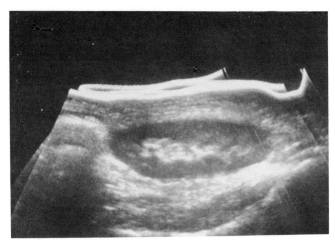 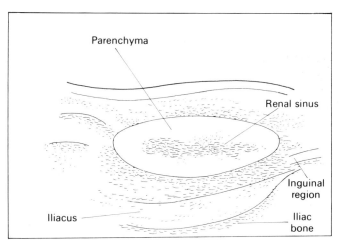

Fig. 8.29 Longitudinal section in the long axis of an established transplant lying in the right iliac fossa. The renal outline is smooth and the sinus echoes compact.

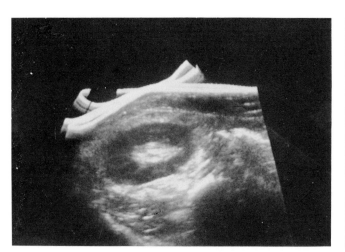 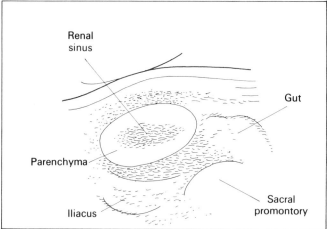

Fig. 8.30 Transverse section of the same kidney as Figure 8.29. The plane of the section is at the level of the anterior superior iliac spine. This plane is used as it is easily repeatable for serial assessment of transplant size.

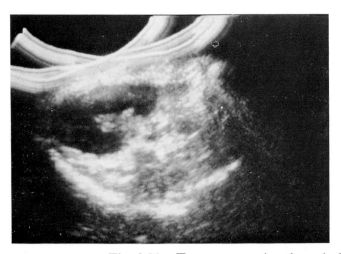 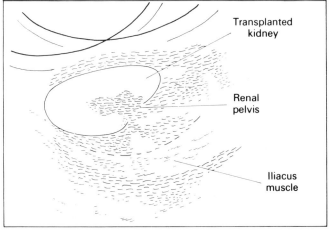

Fig. 8.31 Transverse section through the renal pelvis with the pelvis in a medial position.

THE KIDNEYS 113

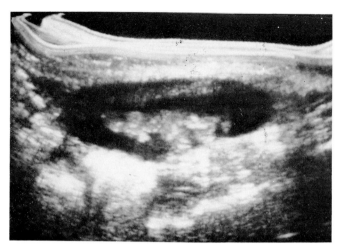 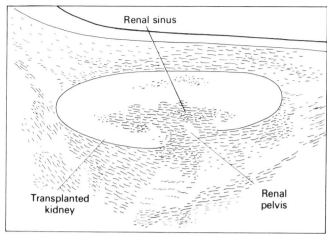

Fig. 8.32 Longitudinal section through the renal pelvis which is in a posterior position in this transplant. The renal pelvis is slightly distended.

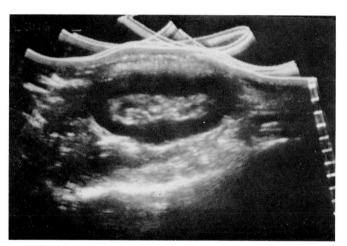 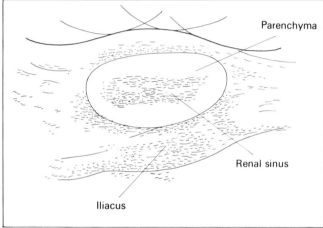

Fig. 8.33 Longitudinal section through the renal sinus with slight separation of the echoes. This is due to distension of the pelvis and is seen in normal kidneys.

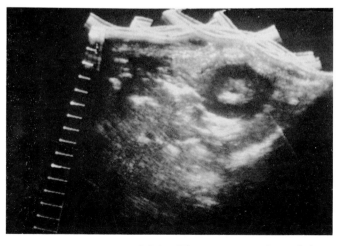 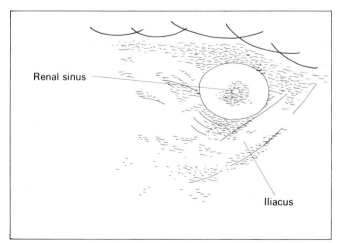

Fig. 8.34 Transverse section of the same kidney as Figure 8.33 with slight distension of the renal pelvis separating the renal sinus echoes.

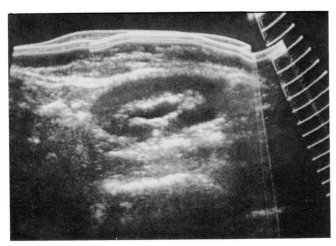 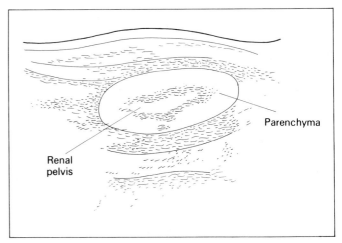

Fig. 8.35 Longitudinal section. The renal pelvis is significantly distended, however renal function was good and there was no evidence of ureteric obstruction or reflux.

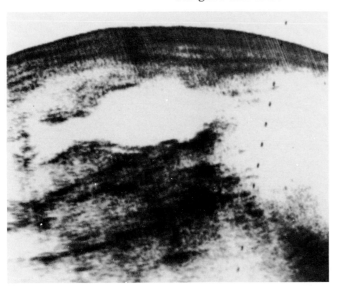 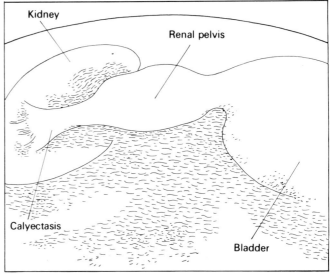

Fig. 8.36 Transverse section. The renal pelvis and calyces are dilated when the bladder is distended. When the bladder was empty there was no evidence of hydronephrosis. In this subject the transient dilation was due to high insertion of the ureter.

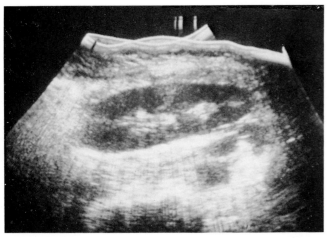 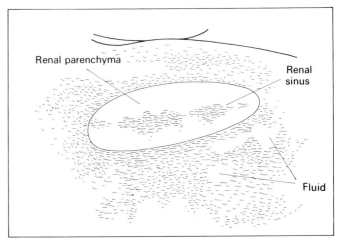

Fig. 8.37 Longitudinal section. Recently transplanted kidney with small fluid collections behind the lower pole which absorbed spontaneously. Small collections are usually of no clinical significance.

THE KIDNEYS 115

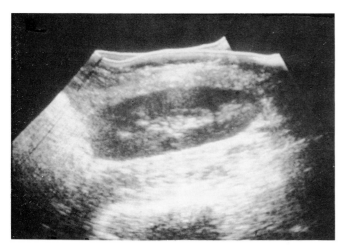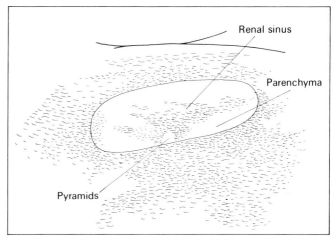

Fig. 8.38 Longitudinal section through the long axis of a recently transplanted kidney. Textural detail is normal with highly reflective sinus structures and a lower density parenchyma.

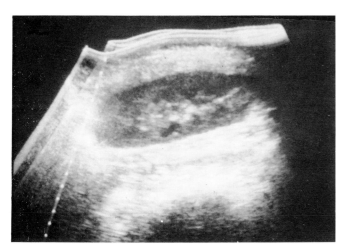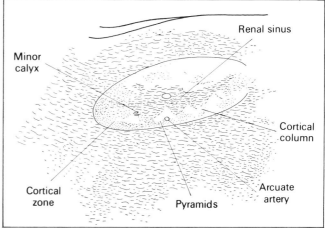

Fig. 8.39 Longitudinal section 1 cm lateral to Figure 8.38. The renal sinus structures are strongly reflective, the cortex and cortical columns are homogeneous and of moderate density, and the pyramids are relatively echo-free.

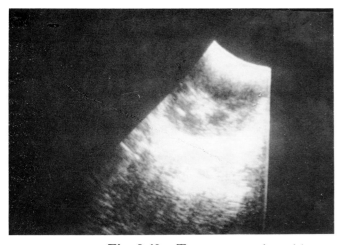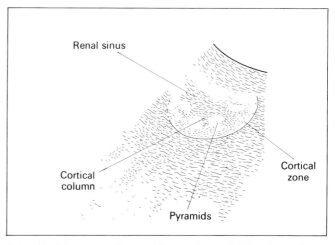

Fig. 8.40 Transverse section with textural detail of the parenchyma. (Scan taken at high gain setting.)

116 ULTRASONIC SECTIONAL ANATOMY

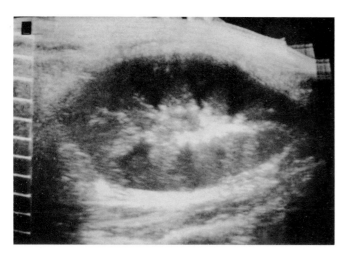 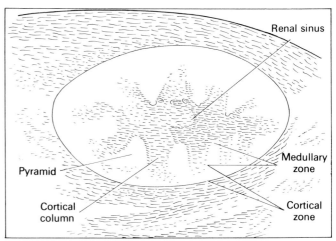

Fig. 8.41 Longitudinal section with the image size enlarged. The renal sinus structures are highly reflective. The cortex and cortical columns are clearly differentiated from the low-density pyramids.

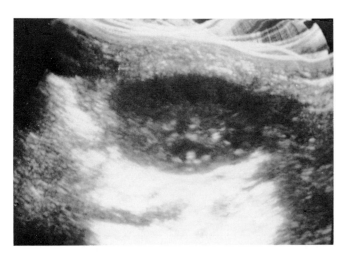 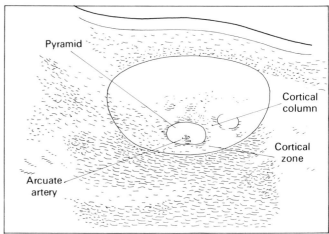

Fig. 8.42 Enlarged image of arcuate vessels which lie in the boundary zone between the cortical and medullary substance.

9 THE VISCERA OF THE LOWER ABDOMEN AND PELVIS

THE URINARY BLADDER, URETER, SEMINAL VESICLES AND PROSTATE

The bladder

In the child the *bladder* is an abdominal organ even when empty. It begins to enter the enlarging pelvis at six years of age, but is not entirely a pelvic organ until after puberty. In the adult the bladder lies in the anterior inferior part of the lesser pelvis. Its superior surface is covered with peritoneum which is reflected posteriorly at the junction of the body and cervix of the uterus in the female, forming the uterovesical pouch, or over the superior surface of the deferent ducts in the male. The bladder lies relatively freely in the surrounding extraperitoneal tissues except at its neck where it is held firmly by ligaments. It is free therefore to expand superiorly.

The bladder is pyramidal in shape with the apex anteriorly, continuous with the median umbilical ligament. The base faces postero-inferiorly, in the female it is in contact with the cervix and upper vagina and in the male with the seminal vesicles and the ampullae of the deferent ducts. The *ureters* enter the bladder at the lateral ends of the bladder base. The bladder also has two inferolateral surfaces and superior surface.

The bladder is examined either by direct contact scanning through the abdominal wall or by radial scanning systems, with the transducer in the rectum or inserted transurethrally directly into the bladder. To obtain maximum diagnostic information the bladder should be fully distended.

The *empty bladder* cannot be defined on the pelvic scan. The partially-filled bladder has a rounded outline and lies in a relatively deep position in the pelvis. As it fills it rises into the lower abdomen assuming an ovoid shape with its long axis directed upwards and forwards. On low transverse sections the bladder is quadrilateral in shape conforming to the boundaries of the lesser pelvis; on higher scans it is more ovoid in contour.

Urine at normal sensitivity settings is echo-free though reverberation artefacts are commonly encountered. The *internal surface* of the normal distended bladder is smooth and highly echogenic, the *outer surface* of the wall is not so clearly defined and wall thickness may only be clearly demonstrated in slim subjects employing high-frequency transducers.

The *ureter* in the adult is approximately 25 cm in length and 5 mm wide. It descends almost vertically on the psoas major muscles along the line of the tips of the lumbar transverse processes. At its mid-point it lies on the origin of the external iliac arteries and then enters the lesser pelvis in front of the internal iliac artery and runs postero-inferiorly deep to the peritoneum of the lateral pelvic wall. In the *female* it passes postero-inferior to the ovary. It then curves medially above the levator ani and in the female it runs beside the lateral vaginal fornix inferior to the broad ligament and the uterine artery, turns superiorly into the broad ligament to reach the bladder base, where it passes obliquely through the wall to open into the bladder at the superiolateral aspect of the trigone. In the *male* it follows a similar course remaining in contact with the peritoneum until it passes behind the ductus deferens to reach the bladder base at the same point as in the female.

The normal ureter is not commonly identified on ultrasonic sections though it may occasionally be seen in the pelvis as it passes behind the ovary in the female, and in its course in the lesser pelvis.

The prostate

The *prostate* is a fibromuscular glandular organ surrounding the bladder neck and proximal portion of the male urethra. It resembles a compressed inverted cone, measuring approximately 3.0 cm by 3.5 cm. The prostate has a rounded anterior surface which lies behind the lower part of the symphysis pubis, two inferolateral surfaces and a posterior surface which is nearly flat lying anterior to the rectum. It has a thin capsule and a loose sheath of visceral pelvic fascia. It consists of two portions, an anterior group of glands associated with the urethra and a posterior portion of a more fibromuscular character. It is divided anatomically into a posterior *median lobe* and two *lateral lobes*, arbitrarily divided by the urethra.

The normal prostate contains low-level echoes and a line of central echoes represent the peri-urethral glands.

Imaging the prostate is not easy as it lies posterior to the symphysis. Using abdominal contact scanning the prostate can be examined through the full bladder, transverse sections being taken with the transducer angled in a caudad plane. Perineal scanning has also been suggested but is not generally employed. These conventional scanning methods provide incomplete images and specially designed radial scanning systems, with the transducer inserted in the rectum and rotating through an angle of 360°, are used to obtain high quality images of outline and soft tissue detail in the horizontal plane (Watanabe et al 1974).

In these sections the normal prostate has a semilunar outline with the A.P. diameter far shorter than the transverse diameter. The shape is symmetrical and the capsule is smooth and thin. The internal echoes are low-level and homogeneous with the urethra and periurethral gland structures defined.

Seminal vesicles

The *seminal vesicles* are paired tubular convoluted structures lying behind the base of the bladder, above the level of the prostate and anterior to the rectum. The tube is coiled upon itself to form a piriform structure. They fuse medially with the *ampulla of the vasa*, forming the ejaculatory ducts.

The seminal vesicles can be seen on sections taken through the full bladder in both the sagittal and horizontal planes. They are also outlined by radial scanning on high sections at the level of the bladder base. On longitudinal sections they appear as a superior extension of the prostate. On transverse sections they have a moustache-like appearance.

The ampulla can sometimes be seen as a tubular anechoic structure located on the posterior superior surface of the bladder and orientated in a longitudinal direction. The vasa deferentia are too small to be seen by ultrasound.

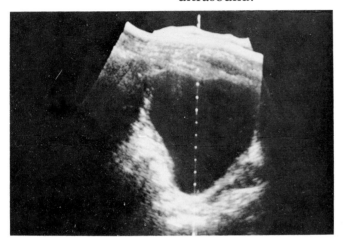
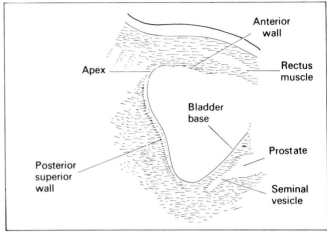

Fig. 9.1 Sagittal section of normal male bladder, prostate and seminal vesicle.

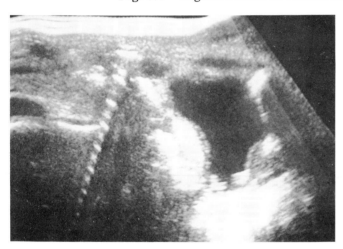
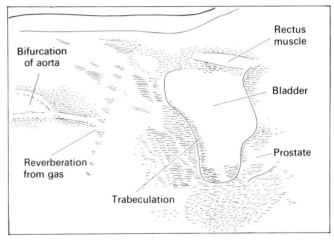

Fig. 9.2 Sagittal section—There is slight prostatic hypertrophy and bladder trabeculation. The subject had no related symptoms.

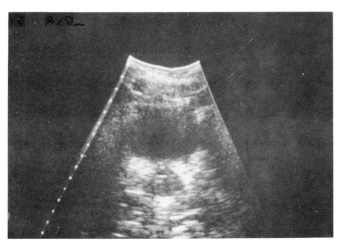
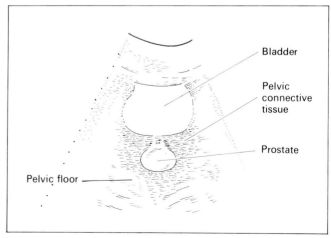

Fig. 9.3 Transverse section with caudad angulation of the plane. A normal prostate is outlined below the bladder.

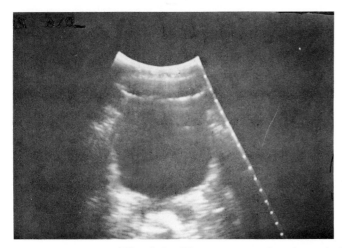 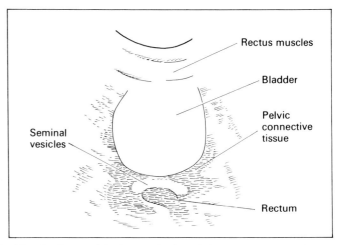

Fig. 9.4 Transverse section through the bladder base with the seminal vesicles outlined anterior to the rectum.

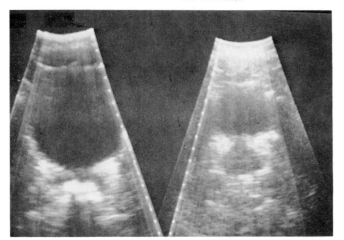 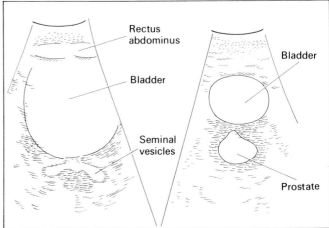

Fig. 9.5 Transverse sector scans in the same subject. The left section is through the bladder base and seminal vesicles. The plane of the section on the right is more caudad and outlines the prostate below the base of the bladder.

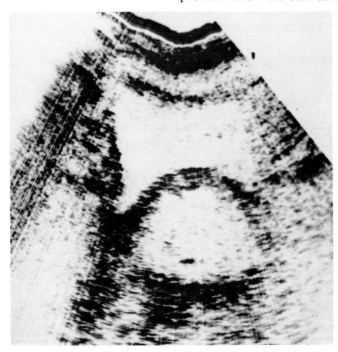 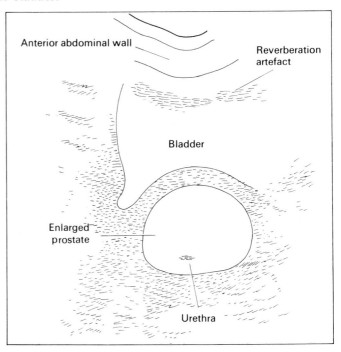

Fig. 9.6 Transverse section. Benign prostatic hypertrophy in an asymptomatic patient. The prostatic urethra is displayed posteriorly.

THE VISCERA OF THE LOWER ABDOMEN AND PELVIS 121

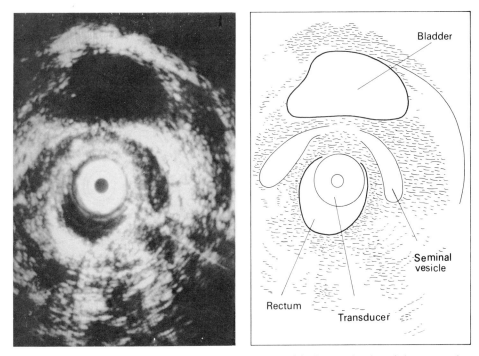

Fig. 9.7 Radial rectal scan at the level of the bladder base with the seminal vesicles anterolateral to the rectum. They fuse centrally with the ampulla of the vasa. The rectum is distended by a water-filled 'balloon'.

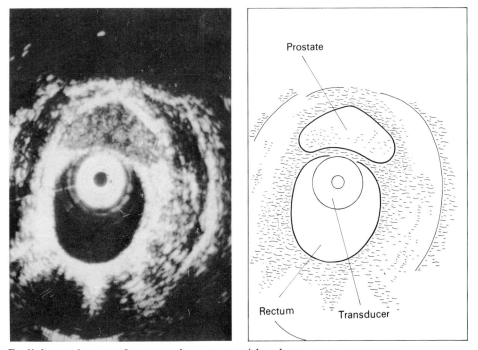

Fig. 9.8 Radial rectal scan of a normal prostate with a homogeneous texture.

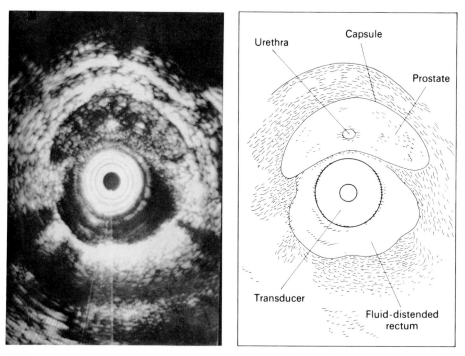

Fig. 9.9 Radial rectal scan. Mild prostate hypertrophy. The prostate retains its normal shape but is slightly enlarged and the texture is of increased density. The urethra is represented by collection of high-density echoes.

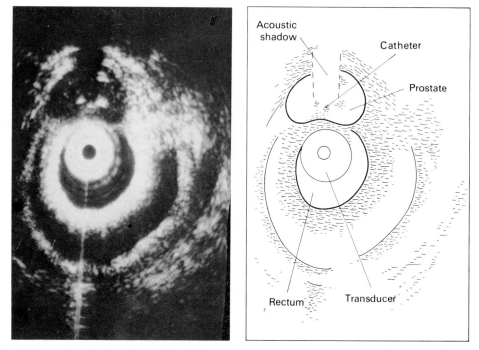

Fig. 9.10 Radial rectal scan. A catheter is in the urethra causing distal shadowing.

THE VISCERA OF THE LOWER ABDOMEN AND PELVIS 123

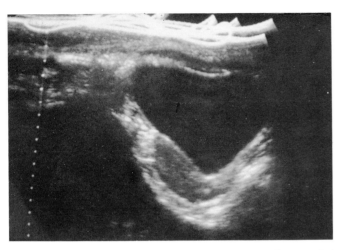 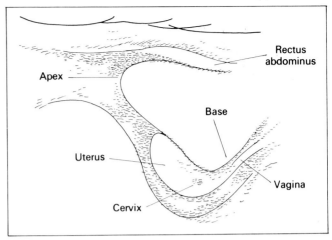

Fig. 9.11 Sagittal section of normal female bladder with the uterus, cervix and vagina outlined posteriorly.

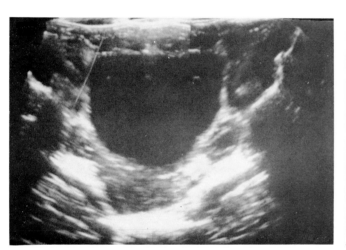 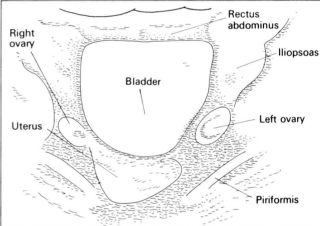

Fig. 9.12 Transverse section through the lesser pelvis. The bladder outline is quadrilateral and the bladder wall thickness defined. Displacement of the bladder wall by the pelvic viscera is commonly encountered. Note the density of the pelvic connective tissue. There is a mature follicle in the left ovary.

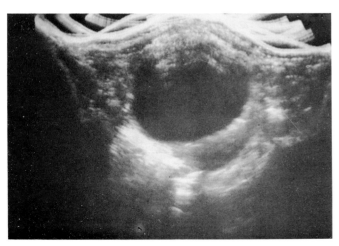 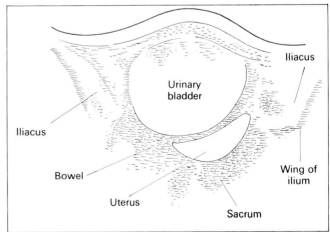

Fig. 9.13 Transverse section through the greater pelvis. The bladder is rounder in outline.

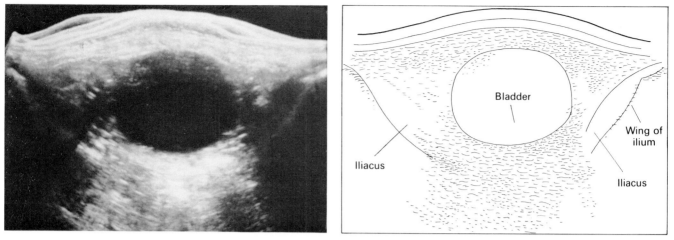

Fig. 9.14 Transverse section through the apex of the bladder. Compare the shape of the bladder in this figure with the shape of the bladder in Figure 9.12.

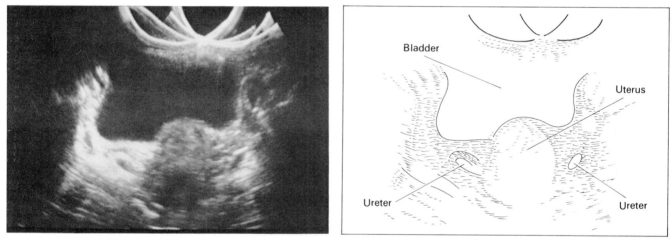

Fig. 9.15 Transverse section. The ureters are outlined as they curve medially towards the trigone. The uterus is bulky. There was no evidence of hydronephrosis or ureteric obstruction.

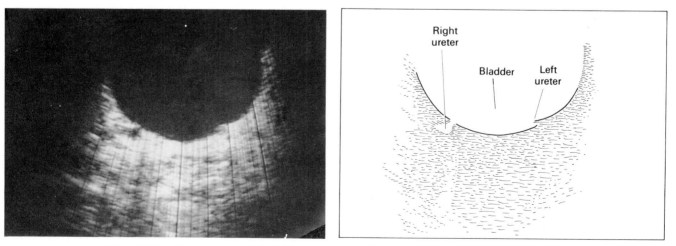

Fig. 9.16 Transverse section. The ureteric orifices are seen as they enter the bladder at the superior lateral aspect of the trigone.

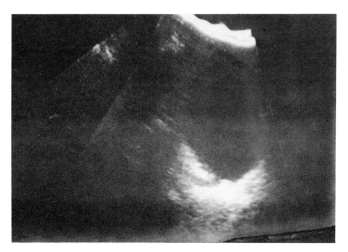 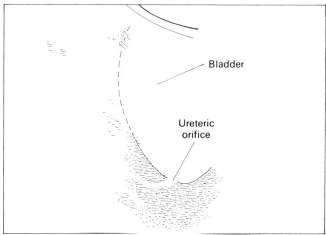

Fig. 9.17 Sagittal section. The lower end of the ureter and the ureteric orifice are seen posteriorly.

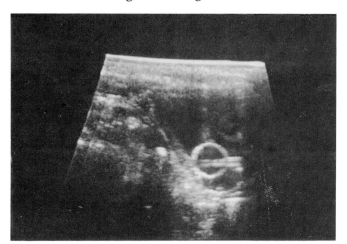 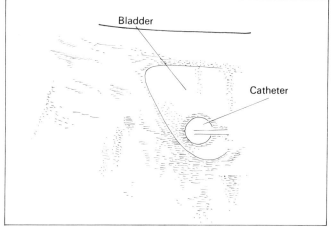

Fig. 9.18 Sagittal section. Bladder with self-retaining catheter. The fluid-distended bulb of the catheter is easily defined and may be used as a method of identifying the bladder during pelvic examination.

THE SCROTUM AND PENIS

The loose skin of the *scrotal sac* is adapted to house the testes, epididymides and spermatic cords. In the median plane there is a ridge or raphe which indicates the embryological line of fusion of the two halves of the scrotum. Deep to the skin is the superficial fascia which contains the dartos muscle. The dartos layer forms a septum between the testes. Each testis is also covered by external spermatic fascia, cremasteric fascia and internal spermatic fascia. In addition the testis is invaginated into the posterior wall of the serous sac—the tunica vaginalis.

The *testis* is oval measuring approximately 4 cm by 2.5 cm by 2.0 cm and is covered by the dense fibrous tunica albuginea. The rete testis is a network of efferent ductules at the upper pole which lead to the head of the epididymis. The *epididymis* overlies the superior and posterolateral aspect of the testis and is divided into head, body and tail. The tail is continuous with the *ductus deferens* which ascends along the medial aspect of the epididymis posterior to the testis in the spermatic cord to the deep inguinal ring.

The scrotum can be scanned immersed in water, covered by a water bath or by direct contact scanning. The open water tank technique is described in the following section using the U.I. Octoson. Examination of the scrotum may be carried out with conventional scanners. The scrotum is supported on a towel and may be scanned in the transverse plane without stabilisation. For longitudinal scans the scrotum is stabilised with the epididymis aligned behind the testis. Recently real-time 'small part' high resolution scanning systems have become available which have small fields of view but provide remarkable detail. The glandular element of the testes normally has a granular texture of medium echogenicity (Sample et al 1978). The epididymis is more echogenic than the testes and sometimes the epididymis or ductus deferens can be resolved.

The *penis* contains two *corpora cavernosa* bound together, side by side, by the tunica albuginea which forms an incomplete septum between them. Their distal ends are embedded in the glans penis and proximally they are attached to the ischiopubic ramus. The *corpora spongiosum* lies between and inferior to the corpora cavernosa, and expands proximally to form the bulb of the penis. The urethra in the male extends from the neck of the bladder to the tip of the glans and is approximately 20 cm in length. It initially traverses the prostate, then pierces the urogenital diaphragm and angles forwards into and passes through the corpus spongiosum to open at the apex of glans. Ultrasonic examination of the penis may be performed with a small parts scanner and by the special technique described in the following section.

EXAMINATION OF THE PENIS AND SCROTUM USING THE U.I. OCTOSON
William J. Garrett and J. Jellins

An open tank technique is necessary for the correct anatomical display of the male external genitalia but open tanks have the disadvantage that the scrotum and surrounding area must be shaved to avoid interference from the hair and there is difficulty in aligning the penis for optimal sections. For routine use a specially designed system is generally required.

With the U.I. Octoson, a standard commercial general purpose echoscope, the patient simply lies prone on a disposable floppy polythene membrane covering a large water tank containing the transducers. The genitalia are lightly compressed so the usual anatomical relations are altered and the epididymis may be difficult to identify. The tissue content of the penis and testis is well displayed and shaving is not necessary as the hair filled with oil is compressed by the membrane.

THE VISCERA OF THE LOWER ABDOMEN AND PELVIS 127

SCANS TAKEN WITH U.I. OCTOSON

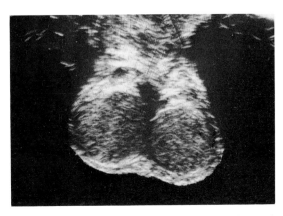 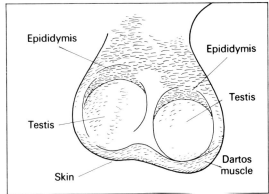

Fig. 9.19 Prone transverse section through the shaved scrotum by the open tank technique. The glandular tissue of the testis is of a granular texture and of medium echogenicity. The epididymis is more echogenic. The dartos is seen as a thin layer of reduced density.

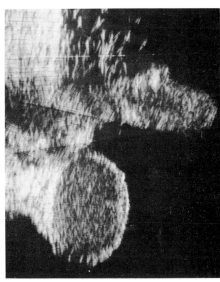 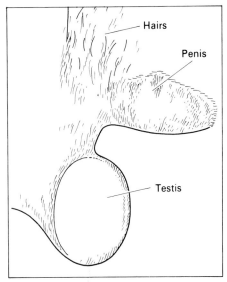

Fig. 9.20 Prone longitudinal section through the left testis, using the open tank technique.

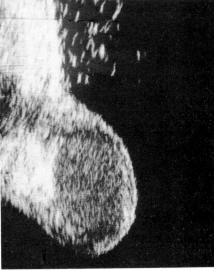 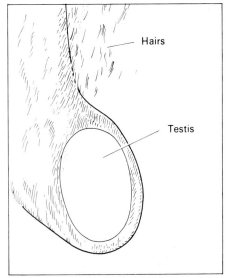

Fig. 9.21 Prone longitudinal section through the right testis, using the open tank technique.

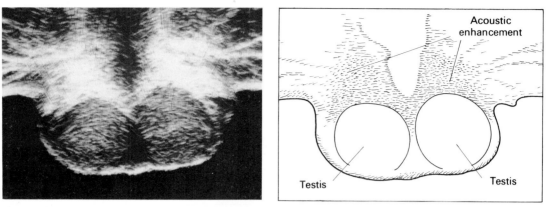

Fig. 9.22 Prone transverse section of the scrotum using a polythene membrane covering the water tank. The scrotum is compressed against the body. Note the enhancement of the echo pattern behind the testes.

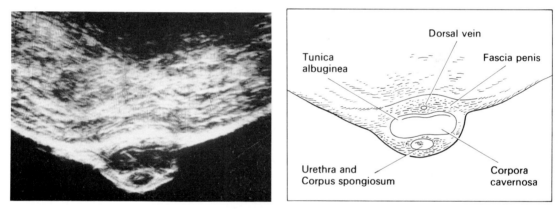

Fig. 9.23 Prone transverse section through the shaft of the penis. (Same technique as 9.22).

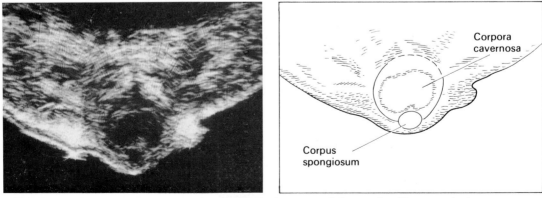

Fig. 9.24 Prone transverse section through the root of the penis. (Same technique as 9.22).

THE VISCERA OF THE LOWER ABDOMEN AND PELVIS 129

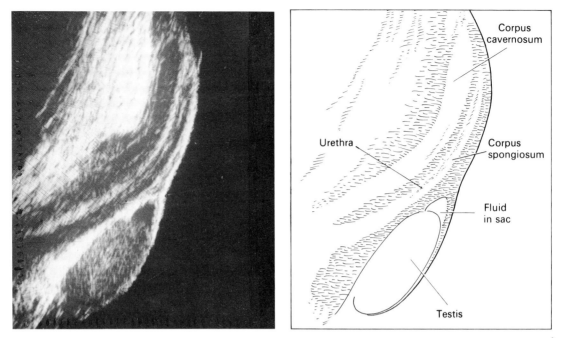

Fig. 9.25 Prone longitudinal section through the penis and left testis. (Same technique as 9.22).

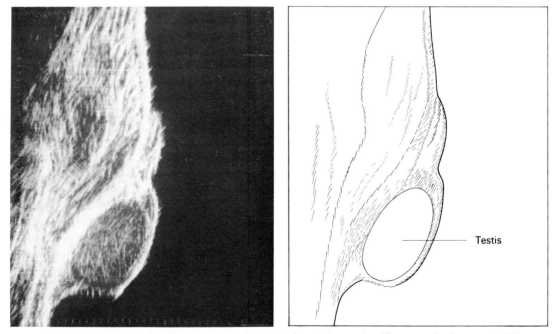

Fig. 9.26 Prone longitudinal section through the right testis. (Same technique as 9.22).

SCANS TAKEN WITH CONVENTIONAL CONTACT SCANNER

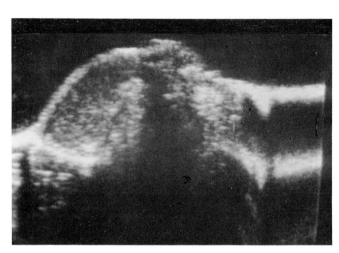 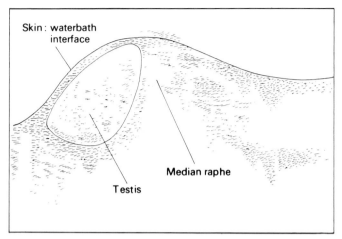

Fig. 9.27 Supine transverse section using the conventional water-bath technique. The testis is granular in texture. The epididymis has not been defined on this section.

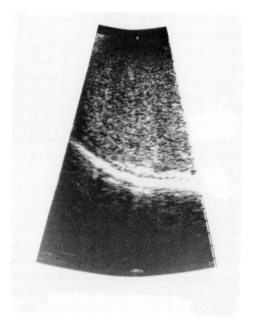 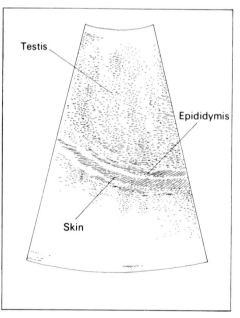

Fig. 9.28A Supine transverse sector scan of the testis and epididymis. The testis is granular in consistency. The epididymis is posterolateral to the testis and is more echogenic.

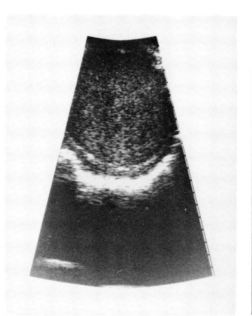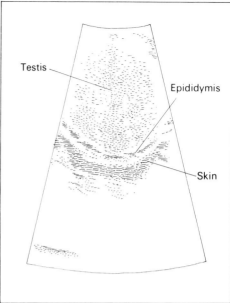

Fig. 9.28B Supine transverse sector scan of the testis with the epididymis defined posteriorly.

THE UTERUS AND OVARIES

The *uterus* lies between the bladder and rectum. It is a thick-walled fibromuscular organ with a narrow lumen or cavity surrounded by *endometrium* which is firmly attached to the myometrium. The endometrium undergoes secretory changes induced by ovarian hormones, proliferative in response to oestrogen and secretory with progesterone. The lumen is continuous with the fallopian tubes and cervical canal.

The uterus is pear-shaped, measuring approximately 8 cm in length in non-pregnant women. It has a fundus, a body, an isthmus and cervix. The cervix is cylindrical (2.5 cm) and is inserted into the vagina through the uppermost part of its anterior wall. The body of the uterus expands superiorly to the fundus (5 cm x 2.5 cm) in the free edge of the broad ligament.

The uterine tubes arise from the superior lateral aspect of the uterus, above the round ligaments and the ligament of the ovary. The tubes are approximately 10 cm long and pass laterally in the broad ligament towards the lateral pelvic wall, curving over the lateral aspect of the ovary and expanding into the fringed funnel-shaped infundibulum which opens into the peritoneal cavity adjacent to the ovary.

The uterine fundus is covered by peritoneum except in its lower anterior position where the bladder wall is contiguous with the lower uterine segment. The anterior reflection of peritoneum forms the uterovesical pouch. On the posterior surface the peritoneum extends to the level of the upper vagina to form the recto-uterine pouch of Douglas. Attached to the uterus on either side is the broad ligament, a transverse fold of peritoneum which encloses the body of the uterus and the parametria. The uterus is supported by condensations of endopelvic fat and fibromuscular tissue laterally at the base of the broad ligament—the cardinal ligaments and uterosacral ligaments. The round ligaments of the uterus arise from the tubo-uterine junction and pass laterally to the side wall of the pelvis, curving forwards to the deep inguinal ring.

The uterus overlies the posterior part of the superior surface and the upper part of the bladder base. With the bladder empty the uterus is anteverted at right angles to the vagina and the plane of the superior aperture of the lesser pelvis. As the bladder fills the uterus is pushed back towards the horizontal mid-position plane in line with the vagina. The vagina is a muscular tube measuring 12–13 cm in depth in the adult. With the bladder empty its axis is towards the sacral promontary in the plane of the superior aperture of the lesser pelvis; as the bladder fills the axis becomes more posterior.

The uterus, parametra and vagina are examined through the filled bladder (page 58). As the axis of the uterus is frequently oblique with the uterus located away from the midline, a long-axis section through uterus and vagina is selected by aligning the uterus and vagina on transverse sections and marking appropriate sites.

The normal adult uterus is pear-shaped, and at its largest can be 9 cm in length, 5 cm in width and 3 cm in A.P. diameter (Piironen 1975). The cervix of the prepubertal uterus shows little difference in size from the adult uterus but the body of the uterus is much smaller. After the menopause the uterus rapidly decreases in size and by the time the patient is over 60 the uterus has shrunk by at least 50%. The uterine outline is smooth and its internal texture homogeneous with a central line of increased density from the lumen. During the course of the menstrual cycle the uterus enlarges after ovulation and the central cavity line becomes more prominent with the proliferation of the endometrium. After pregnancy the uterus takes about 7 weeks to involute back to normal size. Central echoes within the cavity are commonplace at this time due to remaining debris and decidual reaction.

The normal *ovary* is a pinkish-white ovoid structure measuring approximately 2 cm x 3 cm x 3 cm, though it can measure up to 5 cm in any one axis. Normal limits of size for prepubertal ovaries are known (Sample 1979).

Once menstruation commences the ovaries enlarge and vary in size depending on the phase of the menstrual cycle. Between them the ovaries usually produce one ovum per menstrual cycle. This develops as the follicle which ruptures approximately mid-cycle, releasing the ovum into the peritoneal cavity. The remaining lining cells of the follicle develop into a corpus luteum which degenerates towards the end of the menstrual cycle into a fibrous scar, the corpus albicans. After ovulation follicle retention cysts may develop with a normal size limit of about 2 cm, however, larger transient cysts may be seen. Less commonly the corpus luteum may persist with cyst formation. If pregnancy occurs the corpus luteum enlarges progressively for some months reaching its maximum size after about 8 to 10 weeks.

The ovary lies on the posterior surface of the broad ligament inferior to the fallopian tube. It lies in the ovarian fossa near the lateral wall of the pelvis and is closely related to the ureter and the internal iliac vessels posteromedially, and the external iliac vessels anterolaterally. It is supported by the infundibulopelvic ligament and the utero-ovarian ligament.

Ultrasonically the ovaries are located in the angle formed by the lateral wall of the full bladder, the iliopsoas muscle and the lateral pelvic wall. They usually lie approximately at the level of the uterine fundus. They are ovoid with a smooth outline and a characteristic texture. Developing and mature follicles can be seen in the ovarian stroma.

THE VISCERA OF THE LOWER ABDOMEN AND PELVIS 133

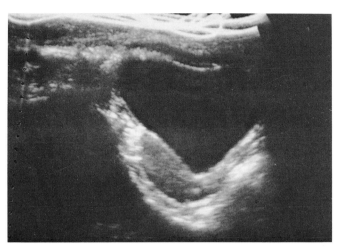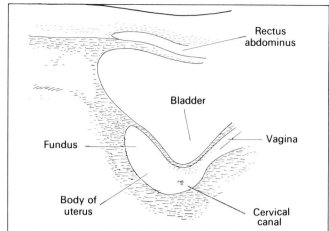

Fig. 9.29 Sagittal section through the long axis of a normal-sized uterus measuring 6 cm in length. The myometrium is homogeneous and the echoes low-level. Note the angle formed between the axes of the uterus and vagina. The echo from the cervical canal is at the level of the external os.

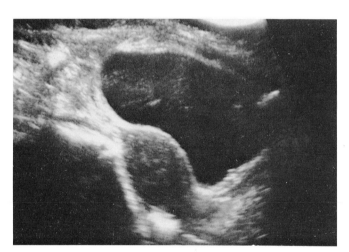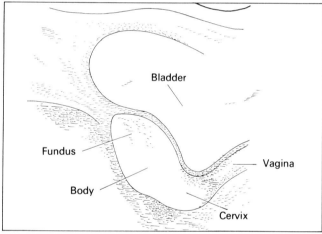

Fig. 9.30 Sagittal section. This uterus is 8 cm in length. There is slight narrowing of uterine outline between the body and the cervix at the level of the isthmus.

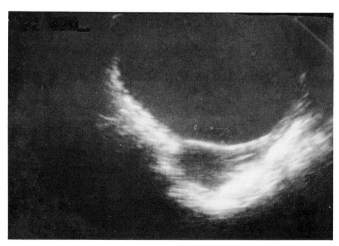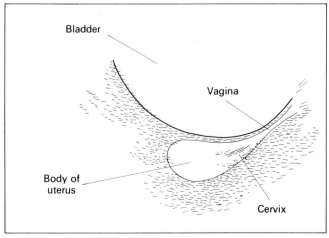

Fig. 9.31 Sagittal section. The uterus has been pushed backwards into a retroverted position by a very full bladder.

134 ULTRASONIC SECTIONAL ANATOMY

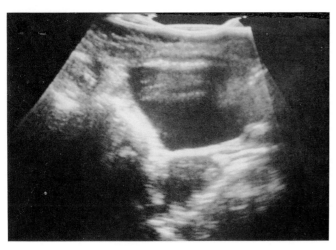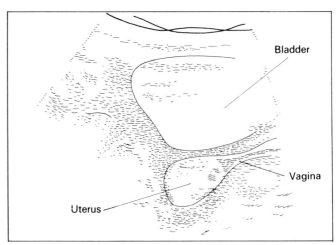

Fig. 9.32 Sagittal section. Retroverted uterus with its axis directed posteriorly. To see detail of the fundus of a retroverted uterus the bladder has to be very distended.

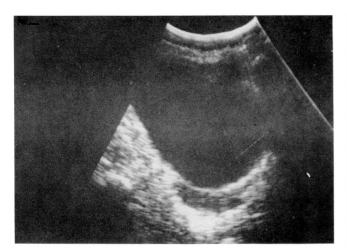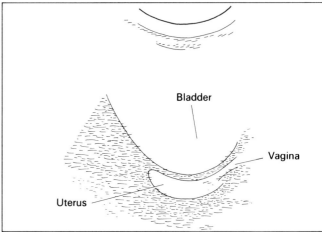

Fig. 9.33 Sagittal section. Senile atrophic uterus. The subject was in her eighth decade.

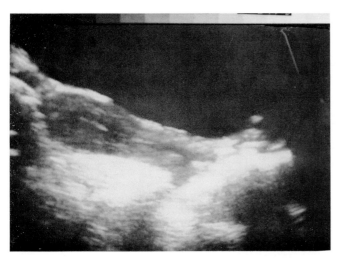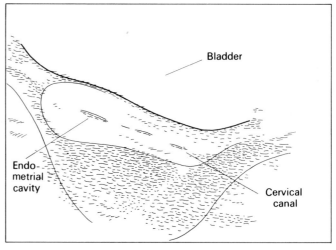

Fig. 9.34 Sagittal section defining the thin cavity of the uterus and the thin cervical canal (scale enlarged to show detail).

THE VISCERA OF THE LOWER ABDOMEN AND PELVIS 135

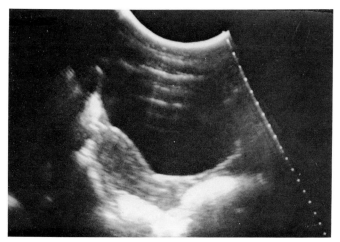 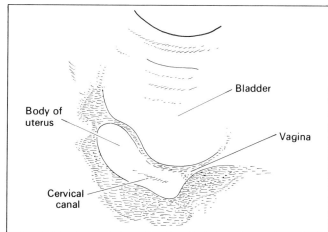

Fig. 9.35 Sagittal section with normal cervical canal. The cavity of the body of the uterus has not been demonstrated in this subject. It is easier to define in the premenstrual period.

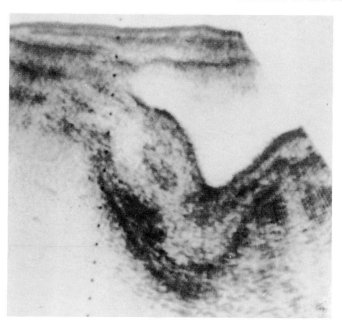 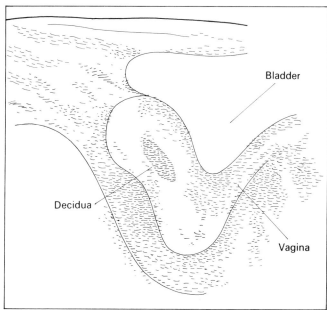

Fig. 9.36 Sagittal section. Thickened endometrium from marked decidual reaction prior to the menstrual period.

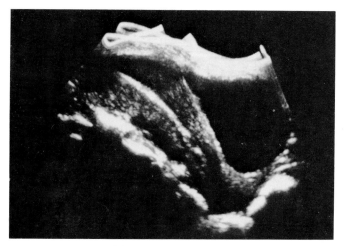 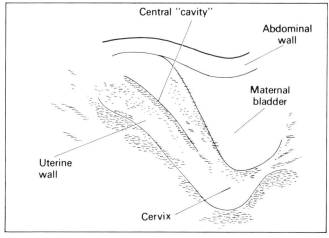

Fig. 9.37 Sagittal section. Bulky puerperal uterus with prominent cavity and thick myometrium.

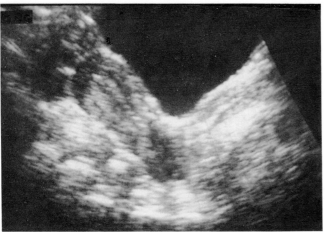 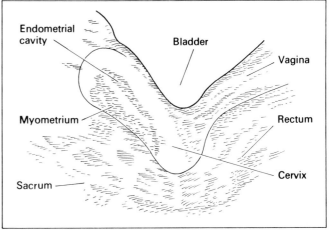

Fig. 9.38A Sagittal section taken at high sensitivity to demonstrate myometrial texture. The cavity is thin. Note the area of reduced density adjacent to the cavity.

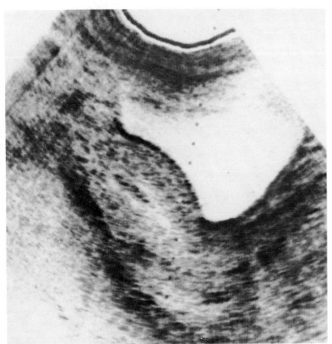 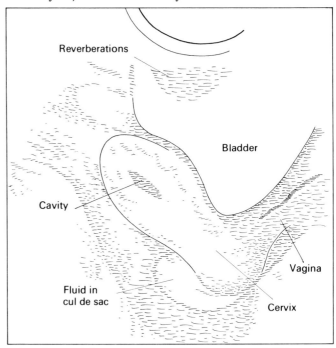

Fig. 9.38B Sagittal section. Myometrial and endometrial texture are well demonstrated. There is a small volume of fluid in the cul-de-sac.

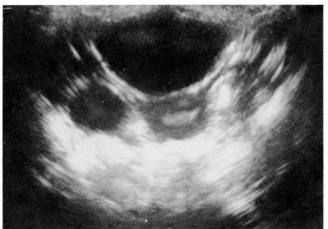 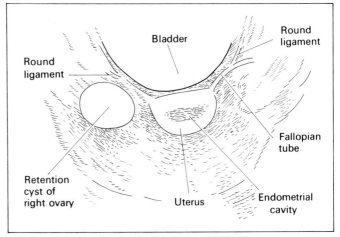

Fig. 9.39 High transverse section through the fundus of the uterus. On the left the fallopian tube is outlined with the round ligament passing upwards towards the deep inguinal ring. On the right the round ligament is seen anterior to a 4 cm diameter retention cyst of the right ovary. There is marked decidual change with thickened endometrium which may be related to oestrogenic activity of the retention cyst.

THE VISCERA OF THE LOWER ABDOMEN AND PELVIS 137

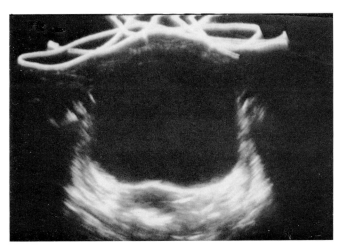
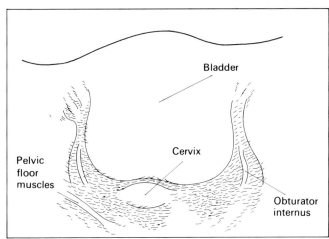

Fig. 9.40 Transverse section at the level of the uterine cervix. To identify the level of the cervix sections are initially made in the sagittal plane. The outline of the cervix is not well-defined laterally on transverse sections, possibly due to the insertion of the transverse ligaments of the cervix.

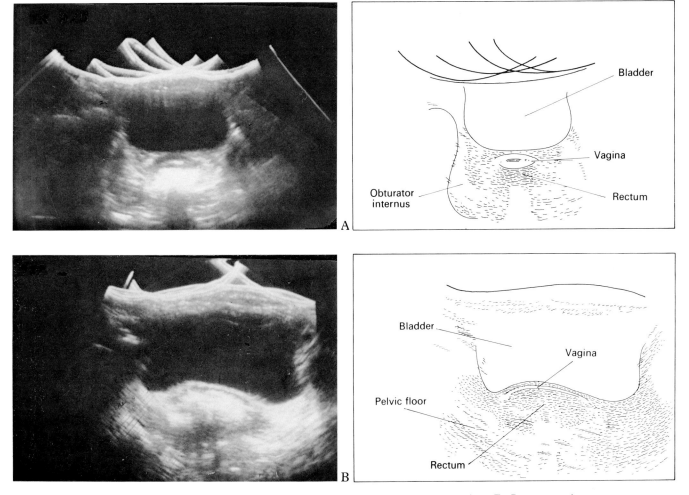

Fig. 9.41 Transverse sections at vaginal level. A. Upper vagina. B. Lower vagina.

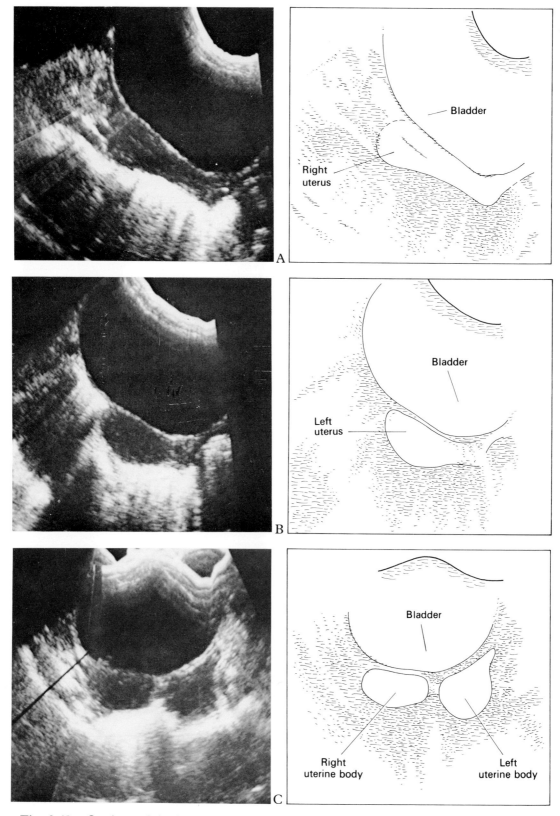

Fig. 9.42 Sections of duplex uterus.
  A. Sagittal section to the right
  B. Sagittal section to the left
  C. Transverse section.
  Only major anomalies can be demonstrated and the diagnosis is best made from transverse scans through the fundal part of the uterus.

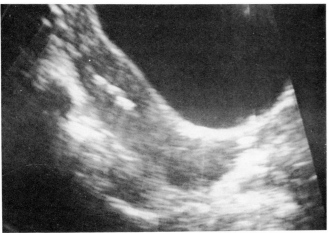 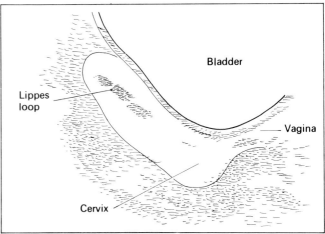

Fig. 9.43A Sagittal section. Highly reflective echoes are present in the upper uterine body. These are due to the presence of an intra-uterine contraceptive device—in this subject a Lippes Loop. These devices usually produce strong echoes though the appearance varies with individual device.

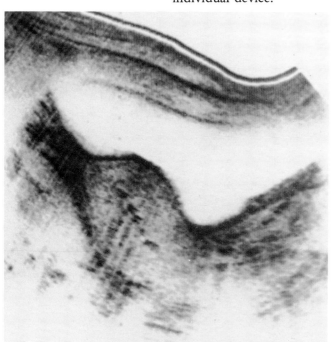 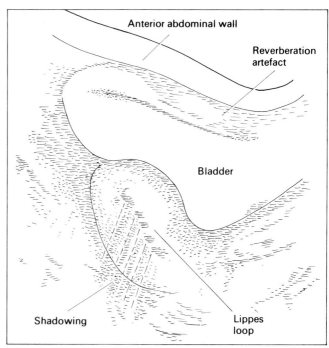

Fig. 9.43B Sagittal section with a Lippes Loop in the uterus. Strong acoustic shadowing is seen, which is characteristic of the Lippes Loop.

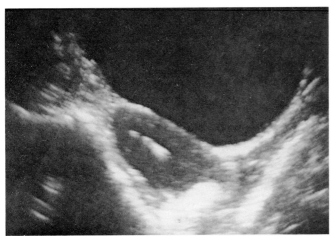 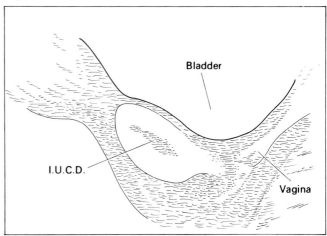

Fig. 9.44 Sagittal section. There is a Copper 7 in the uterus. This device is not so strongly reflective and does not cause acoustic shadowing.

140 ULTRASONIC SECTIONAL ANATOMY

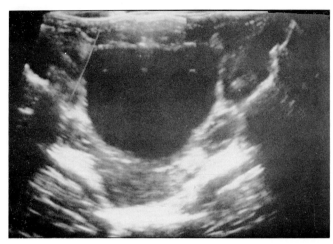 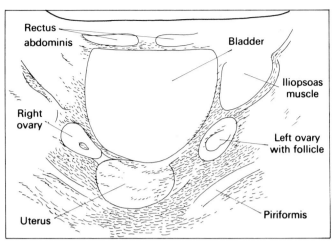

Fig. 9.45A Transverse section through the body of the uterus. Both ovaries are outlined in the angle between the bladder, iliopsoas muscle and the lateral pelvic wall. There is a large follicle in the left ovary and a small follicle in the right ovary.

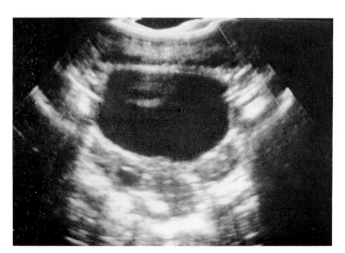 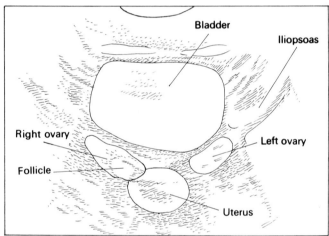

Fig. 9.45B Transverse section. This scan is taken at higher sensitivity so that the texture of the ovarian stroma is demonstrated. The right ovary contains a developing follicle.

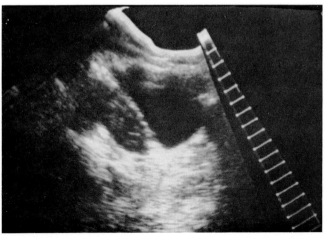 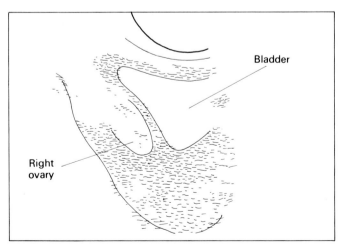

Fig. 9.46A Sagittal section of right ovary. Slight lateral angulation of the scanning plane may be necessary to outline the ovaries on sagittal sections.

THE VISCERA OF THE LOWER ABDOMEN AND PELVIS 141

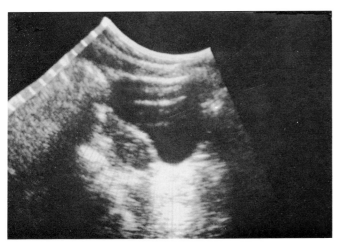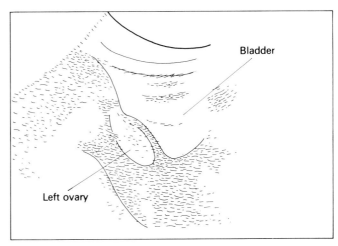

Fig. 9.46B Sagittal section of the left ovary. Slight lateral angulation of the scanning plane may be necessary to outline the ovaries on sagittal sections.

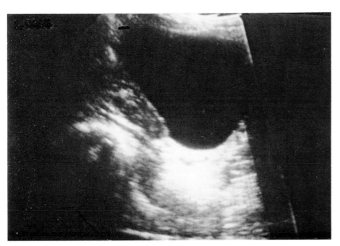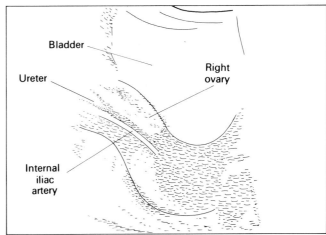

Fig. 9.47A Sagittal section. The right ovary is outlined anterior to the internal iliac artery. This is a useful anatomical landmark for identification of the ovary. The ureter is obliquely sectioned behind the upper pole of the ovary as it passes anterior to the artery.

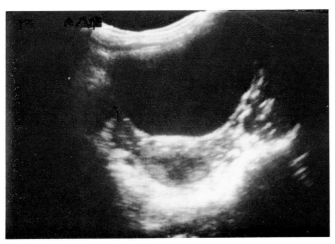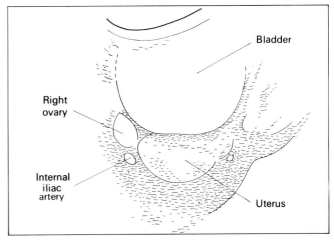

Fig. 9.47B Transverse section with the internal iliac artery behind the right ovary. This section is angled in a cephalad plane.

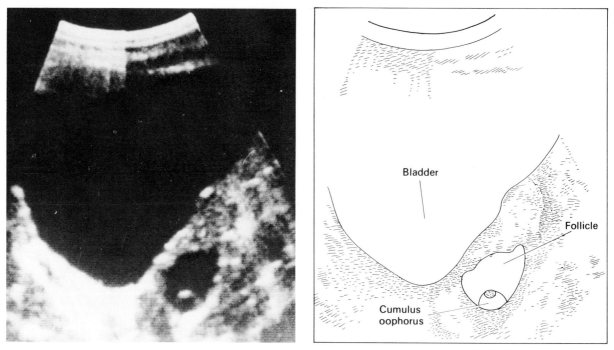

Fig. 9.48 Limited sector scan showing detail of an ovarian follicle with the cumulus oöphorus defined posteriorly.

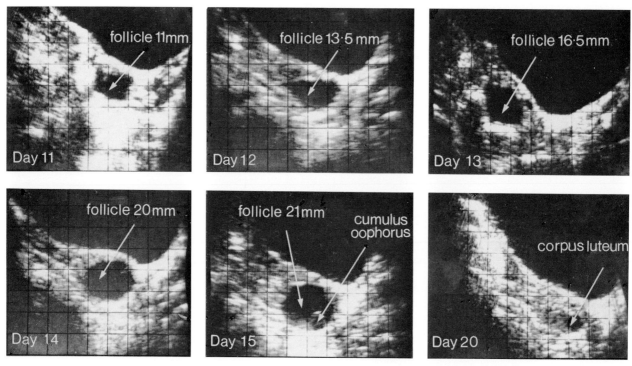

Fig. 9.49 Limited sector scans of the same ovary, monitoring follicular development from the 11th to the 15th day of the menstrual cycle and demonstration of the corpus luteum on the 20th day.

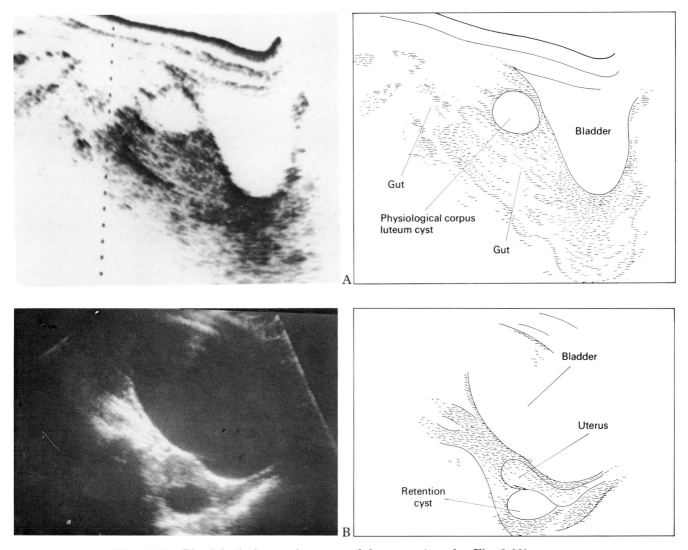

Fig. 9.50 Physiological retention cysts of the ovary (see also Fig. 9.39).
A. Sagittal section. A corpus luteum cyst which spontaneously regressed. This is located in the normal anatomic position of the ovary.
B. Sagittal section. Small retention cyst behind the uterus. The ovary has dropped into the pouch of Douglas.

10 THE GASTRO-INTESTINAL TRACT AND PERITONEAL CAVITY

THE GASTRO–INTESTINAL TRACT

Ultrasound is not an imaging technique that is primarily employed for examination of the stomach and intestines. A detailed discussion of alimentary tract anatomy is therefore not included in this chapter.

The appearance of an individual segment of the gastro-intestinal tract depends significantly on its content. However, certain segments may be identified from their pattern and anatomic position and require to be recognised and differentiated from the other contents of the abdominal and pelvic cavities. The greater and lesser omentum, the mesentery, peritoneal ligaments including the transverse and sigmoid mesocolon are not defined unless grossly infiltrated or unless peritoneal fluid is present.

The intestinal tract is usually distended with varying amounts of air, fluid or more solid faecal material. Air is highly reflective and causes acoustic shadowing, reverberation artefacts and loss of distal information. Barium may cause similar problems (Leopold & Asher 1971, Sarti & Lazere 1978). Fluid-filled segments may be mistaken for peritoneal fluid collections. However with real-time systems movement of the fluid and peristalsis can be recognised and a fluid-distended stomach or small intestine identified. Administration of fluid orally will also assist in the localisation of the stomach and proximal duodenum, or with a fluid collection in the left upper quadrant diagnostic aspiration with a gastric tube will empty the fluid-distended stomach.

Fluid and faeces form pseudomasses which are a particular problem in the pelvis. Observations of the mass during administration of water by rectum—the water enema (Rubin et al 1978)—with real-time equipment will allow discrimination between a pseudomass in which fluid movement can be observed and a true mass of diagnostic significance.

Collapsed gut may be seen as a 'target lesion'. There is a highly reflective central echo from mucus and air with an area of sonolucency peripherally representing the surrounding wall. Outlining the wall, a thin highly reflective interface represents the serosal surface and mesentery. Seen in cross-section this produces the target lesion, longitudinally it is displayed as parallel lines (Lutz & Petzoldt 1976, Sample & Sarti 1978).

THE GASTRO-INTESTINAL TRACT AND PERITONEAL CAVITY 145

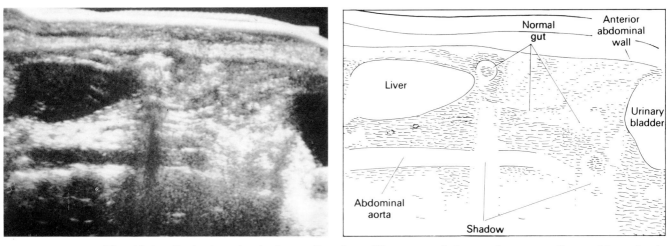

Fig. 10.1 Sagittal section in the median plane. The very varied acoustic pattern of normal intestine is demonstrated.

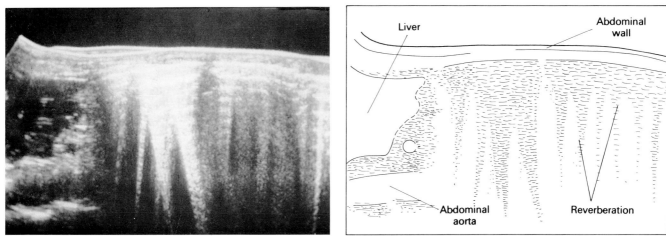

Fig. 10.2 Sagittal section in the median plane in a subject with gross gaseous distension of the small intestine. Marked reverberation artefacts are present and there is loss of distal echo information. Compare with Figure 10.1.

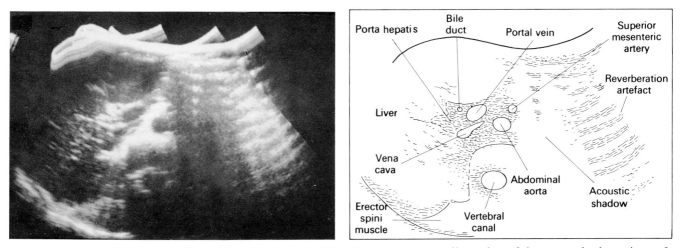

Fig. 10.3 High transverse section. There is gross gaseous distension of the stomach obscuring soft tissue detail in the left upper quadrant.

146 ULTRASONIC SECTIONAL ANATOMY

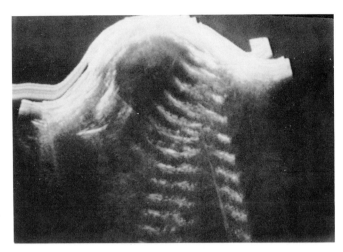 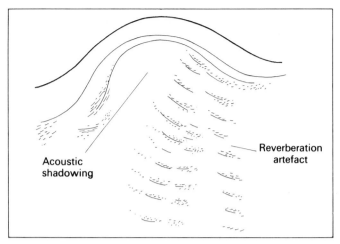

Fig. 10.4 Transverse section at the level of the anterior superior iliac spine. Gaseous distension of the small intestine is causing shadowing and reverberation artefacts with loss of soft tissue detail.

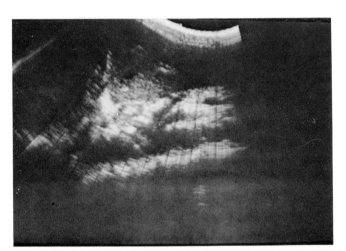 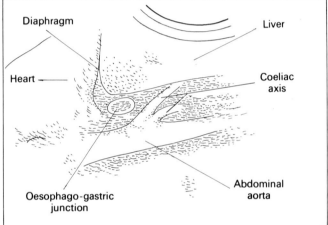

Fig. 10.5 Sagittal section. The oesophagogastric junction is seen anterior to the abdominal aorta. It has the typical 'target lesion' appearance.

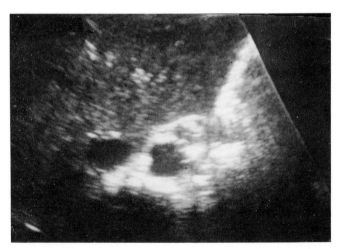 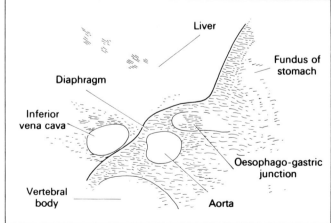

Fig. 10.6 High transverse subxiphisternal section. The oesophagogastric junction is identified anterior to and to the left of the abdominal aorta. The gastric fundus is to the left of the section.

THE GASTRO-INTESTINAL TRACT AND PERITONEAL CAVITY 147

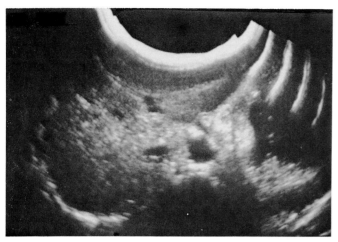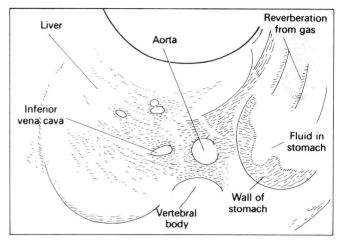

Fig. 10.7 Transverse subxiphisternal section. The stomach is outlined in the left upper quadrant of the abdomen. Fluid with some debris and a gas bubble which lies anteriorly are seen in the stomach. The wall of the stomach is partially defined.

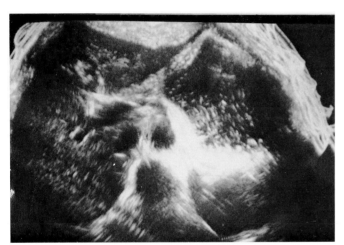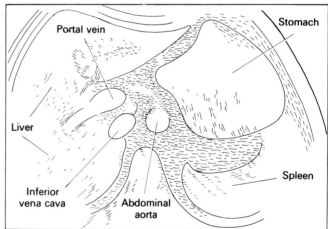

Fig. 10.8 Transverse section in the epigastrium. The stomach is distended with fluid and food debris. This is a typical 'pseudotumour'.

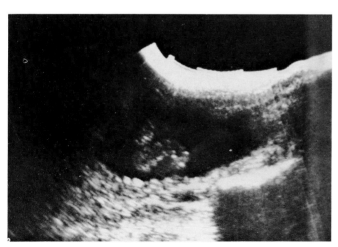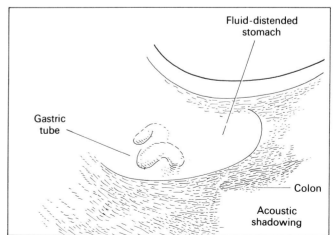

Fig. 10.9 Sagittal section. A gastric tube is seen in the fluid distended stomach. Behind the stomach there is a strongly reflective linear echo with distal shadowing, produced by gas in the splenic flexure of the colon.

148 ULTRASONIC SECTIONAL ANATOMY

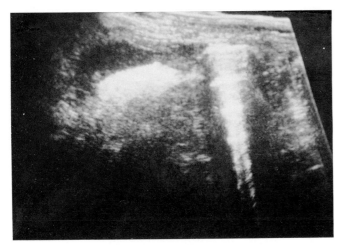 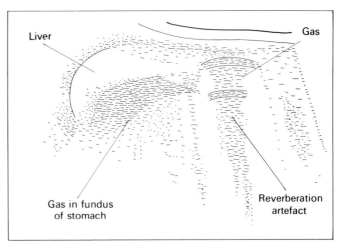

Fig. 10.10 Sagittal section in the left vertical plane. Gas in the fundus and body of the stomach is completely obscuring soft tissue detail. A small section of the left lobe of the liver is seen anteriorly.

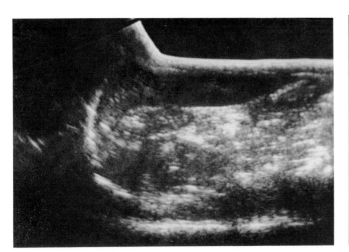 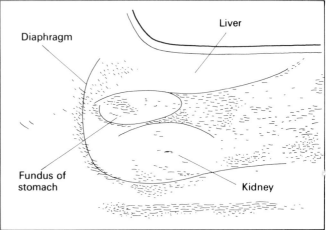

Fig. 10.11A Sagittal section in the left vertical plane. The empty fundus of the stomach is defined behind the left lobe of the liver and anterior to the left kidney. This appearance is typical of a 'pseudotumour'.

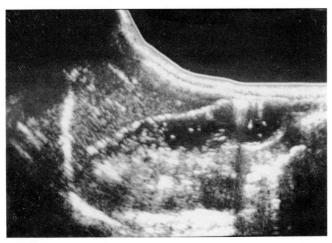 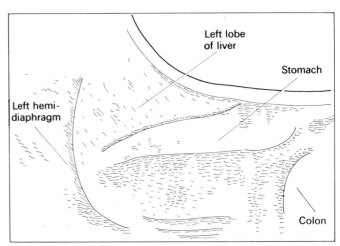

Fig. 10.11B Sagittal section in the left vertical plane. The stomach is distended with fluid and food debris. The colon is behind the inferior border of the stomach below the left kidney.

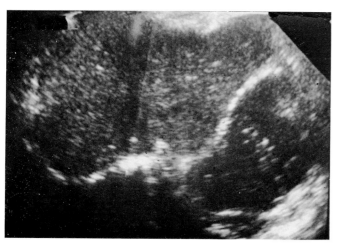 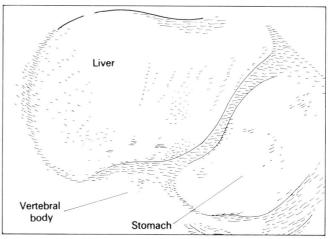

Fig. 10.12A Supine transverse section. Subject prepared for pelvic examination by a water load taken orally. The stomach is distended with fluid and contains food debris.

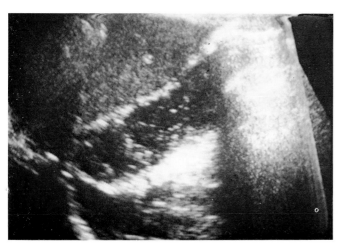 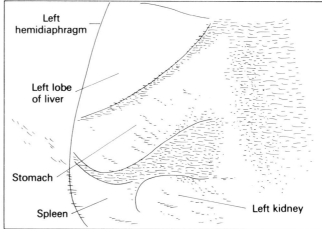

Fig. 10.12B Supine sagittal section. Subject prepared for pelvic examination by a water load taken orally. The stomach is distended with fluid and contains food debris.

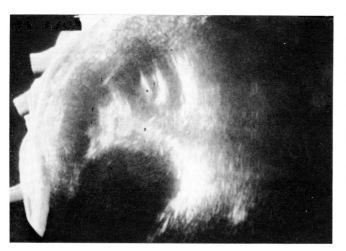 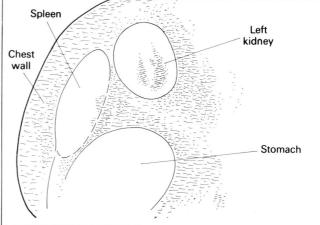

Fig. 10.12C Prone transverse section. Subject prepared for pelvic examination by a water load taken orally. The stomach is distended with fluid and contains food debris.

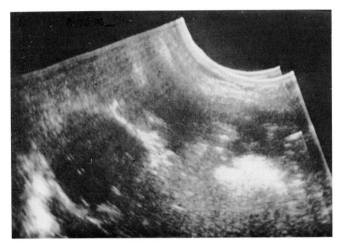 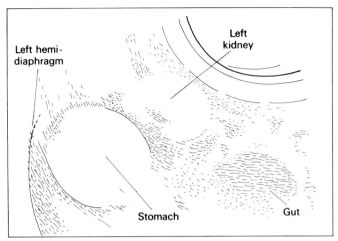

Fig. 10.12D   Prone sagittal section. Subject prepared for pelvic examination by a water load taken orally. The stomach is distended with fluid and contains food debris.

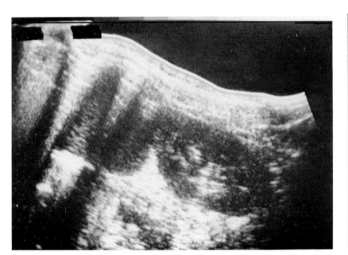 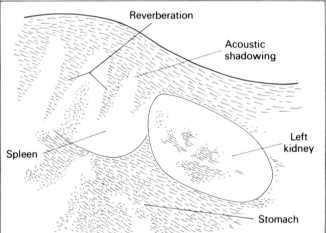

Fig. 10.13   Prone sagittal section. The empty stomach is seen as an ill-defined 'mass' anterior to the spleen and kidney.

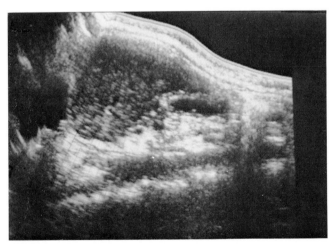 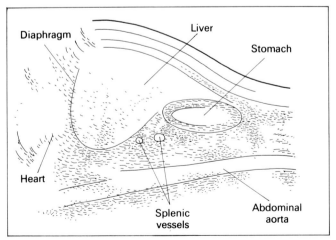

Fig. 10.14   Sagittal section. The narrowing body of the stomach is outlined behind the inferior edge of the left lobe of the liver anterior to the abdominal aorta.

THE GASTRO-INTESTINAL TRACT AND PERITONEAL CAVITY 151

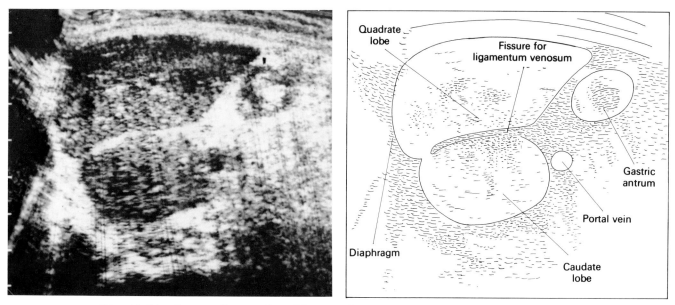

Fig. 10.15 Paramedian section. The gastric antrum is seen below the left lobe of the liver. In this section it is collapsed with the typical 'target lesion' appearance.

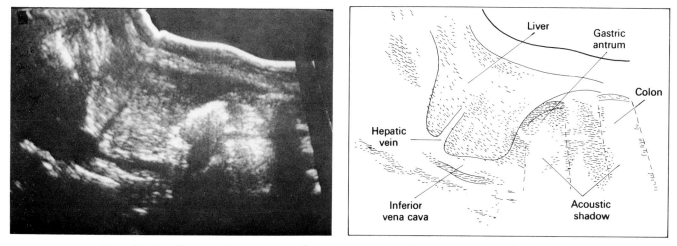

Fig. 10.16 Paramedian section. Gas is present in the antrum obscuring distal detail. Below the stomach the transverse colon, also containing gas, casts a second acoustic shadow.

152 ULTRASONIC SECTIONAL ANATOMY

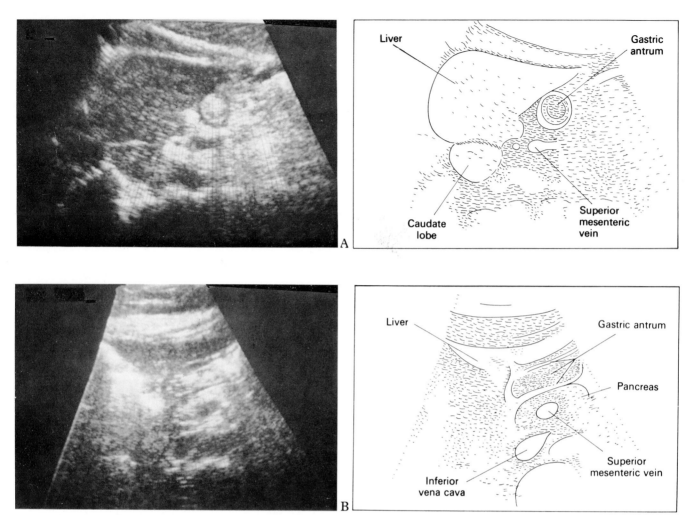

Fig. 10.17 Limited sector scans of the gastric antrum anterior to the pancreas and superior mesenteric vein.
A. Sagittal section with target lesion
B. Transverse section with parallel lines.

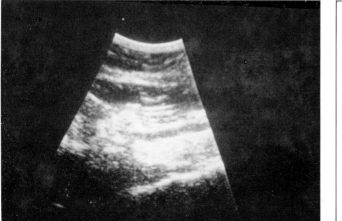 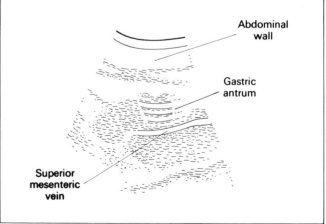

Fig. 10.18 Limited sagittal sector scan in the epigastrium. The gastric antrum is seen as a layer of parallel lines anterior to the superior mesenteric vein.

THE GASTRO-INTESTINAL TRACT AND PERITONEAL CAVITY 153

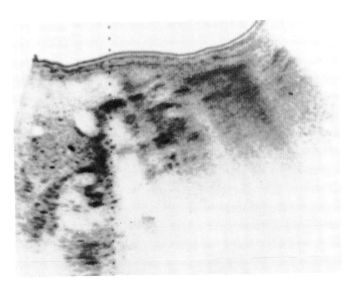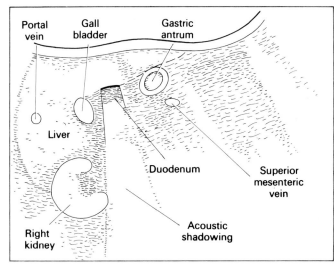

Fig. 10.19 Transverse section in the epigastrium. The superior part of the duodenum (the duodenal cap) contains air and is lying adjacent to the gall bladder. There is acoustic shadowing with loss of distal detail.

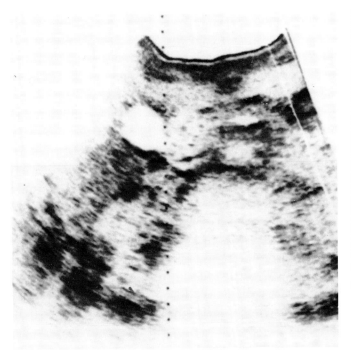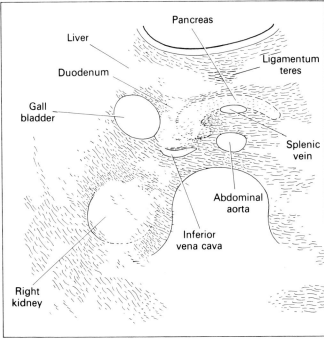

Fig. 10.20 Transverse section in the epigastrium. The duodenum is distended with fluid and lies adjacent to the head of the pancreas.

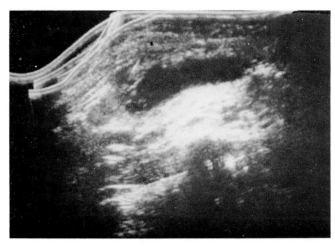 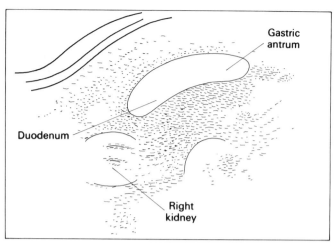

Fig. 10.21 Limited transverse sector scan. The antrum and superior duodenum are grossly distended with fluid. On a static scan this could be mistaken for a peritoneal fluid collection. With a real-time system fluid and debris movement was identified.

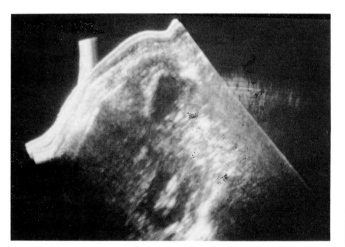 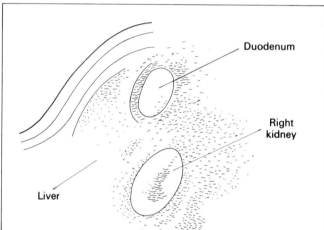

Fig. 10.22 Limited transverse sector scan. The superior part of the duodenum is in an anterior and lateral position and distended with fluid. On this section it could be mistaken for a thick-walled gall bladder.

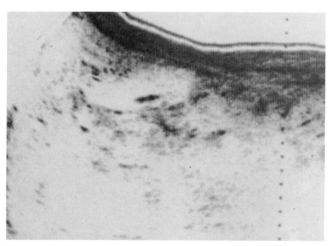 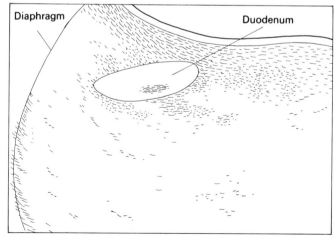

Fig. 10.23 Sagittal section. The superior part of the duodenum is anterior in position. On this section it mimics a gall bladder containing a calculus.

THE GASTRO-INTESTINAL TRACT AND PERITONEAL CAVITY 155

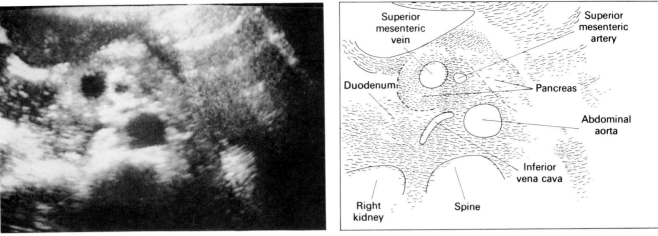

Fig. 10.24A Transverse section. The descending duodenum lies lateral to the head of the pancreas and anterior to the right kidney. It is filled with fluid and debris.

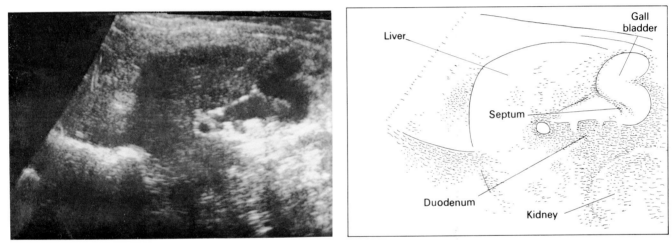

Fig. 10.24B Sagittal section. The descending part of the duodenum is anterior to the kidney and behind the gall bladder.

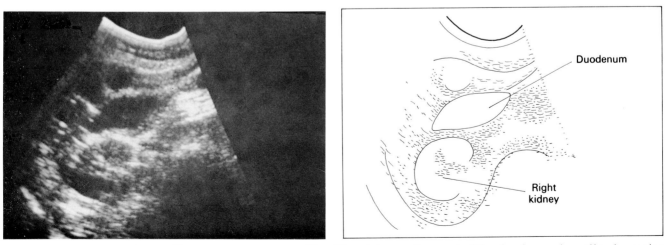

Fig. 10.24C Transverse sector scan. The dilated descending part of the duodenum is outlined anterior to the right kidney.

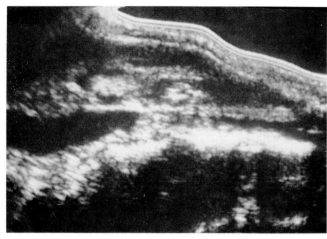 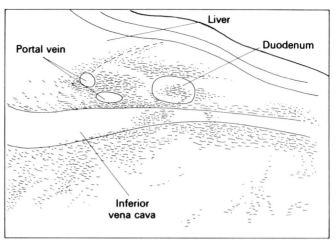

Fig. 10.25 Sagittal section. The horizontal part of the duodenum is outlined anterior to the inferior vena cava below the pancreas. It has a target configuration.

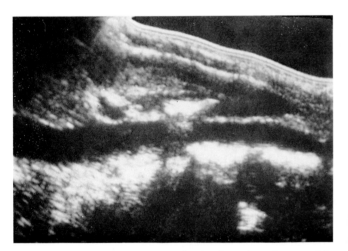 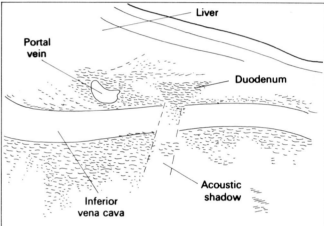

Fig. 10.26 Sagittal section. Same subject as Figure 10.25. Gas is now present in the horizontal part of the duodenum. This scan was recorded shortly after Figure 10.25 and demonstrates the importance of serial observation in identifying segments of gut.

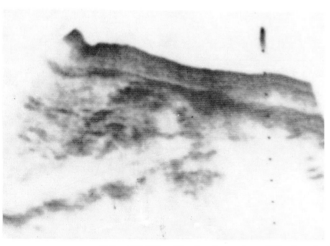 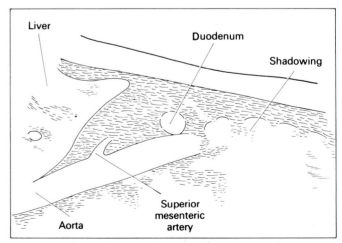

Fig. 10.27 Sagittal section. The horizontal part of the duodenum is anterior to the aorta. Inferiorly there is shadowing from the small intestine.

THE GASTRO-INTESTINAL TRACT AND PERITONEAL CAVITY 157

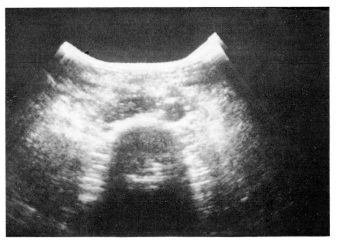 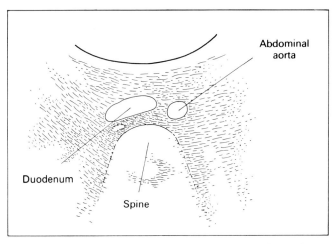

Fig. 10.28 Transverse section above the umbilicus. A short segment of the collapsed horizontal part of the duodenum is seen anterior to the vertebral column.

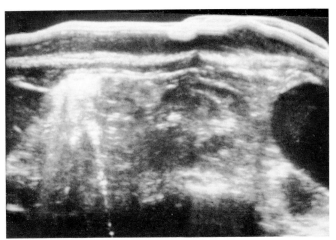 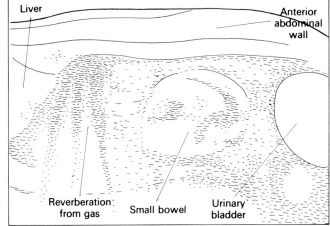

Fig. 10.29 Sagittal section. Fluid-filled loops of small intestine are seen in the lower abdomen. This is a typical 'ultrasonic pseudotumour'.

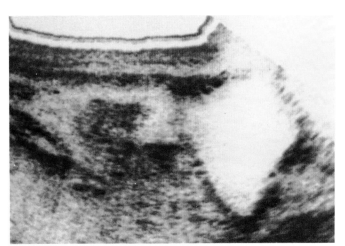 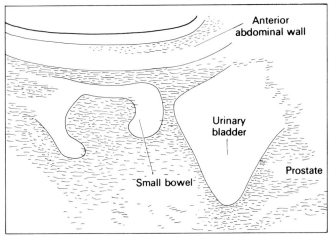

Fig. 10.30 Sagittal section. A distended loop of small intestine lies in the lower abdomen above the bladder.

158 ULTRASONIC SECTIONAL ANATOMY

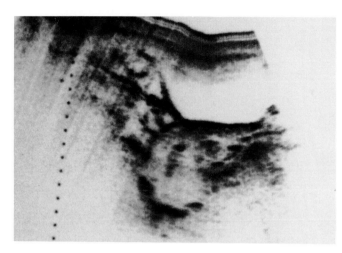 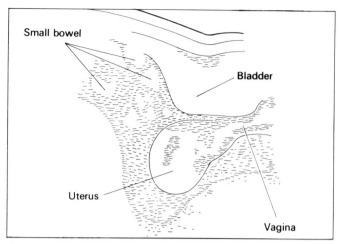

Fig. 10.31 Sagittal section. There are multiple fluid-distended loops of small intestine in the pelvis above the retroverted uterus, extending into the uterovesical pouch.

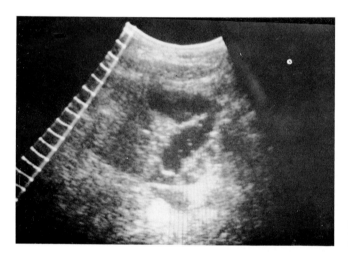 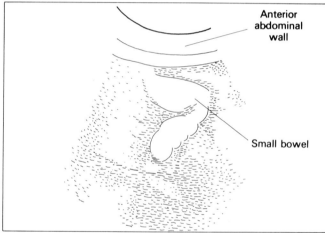

Fig. 10.32 Transverse sector scan in the pelvis. An easily recognisable fluid-distended loop of small intestine, with typical circular folds.

THE GASTRO-INTESTINAL TRACT AND PERITONEAL CAVITY 159

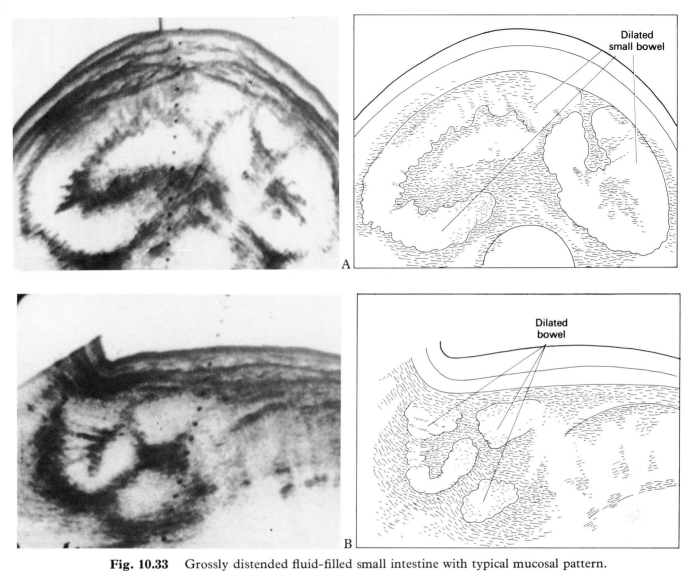

Fig. 10.33 Grossly distended fluid-filled small intestine with typical mucosal pattern.
A. Transverse section
B. Longitudinal section.
Jejunum is on the right and ileum on the left of the abdomen.

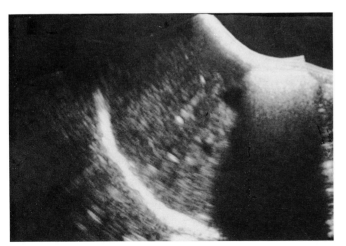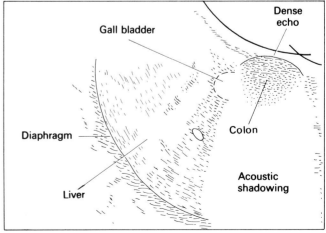

Fig. 10.34 Sagittal section in the right upper quadrant. The hepatic flexure of the colon is distended with gas and lies immediately below and partially obscuring the gall bladder. (The splenic flexure of the colon is seen in Figs. 10.9 and 10.11B.)

160 ULTRASONIC SECTIONAL ANATOMY

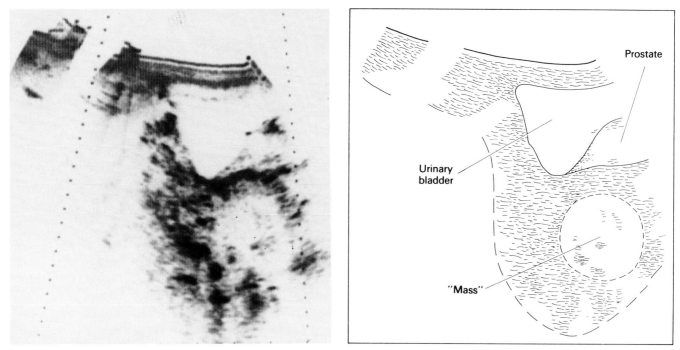

Fig. 10.35A Sagittal section in the pelvis. A mass is seen in the posterior pelvis behind the bladder and prostate.

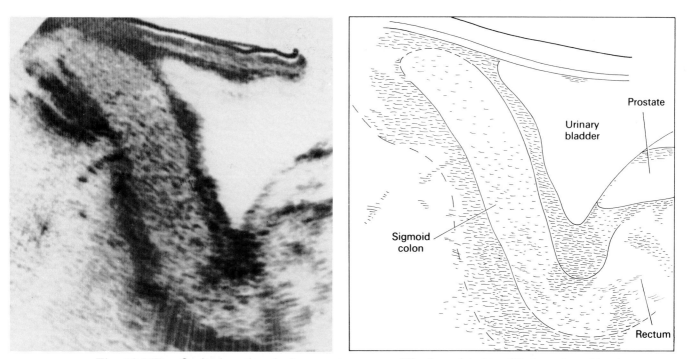

Fig. 10.35B Sagittal section during a water enema. The 'mass' is identified as sigmoid colon.

THE GASTRO-INTESTINAL TRACT AND PERITONEAL CAVITY 161

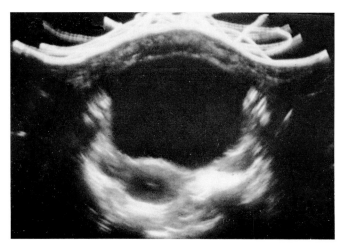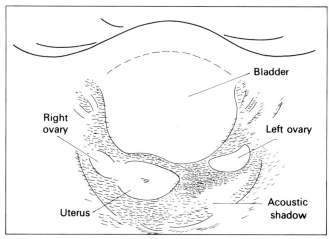

Fig. 10.36 Transverse suprapubic section. There is a strongly reflective echo and acoustic shadowing from gas in the sigmoid colon in the left side of the pelvis.

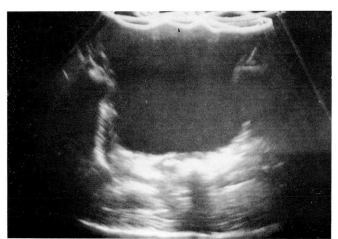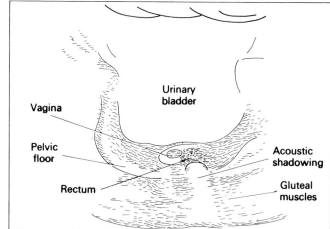

Fig. 10.37 Transverse suprapubic section. Gas is present in the rectum behind the vagina.

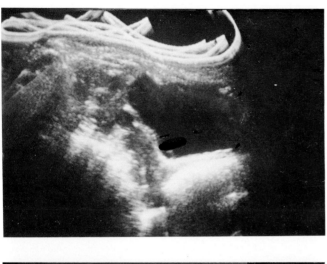
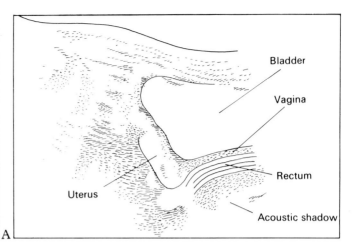
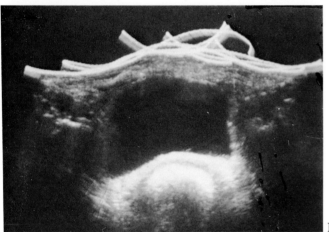
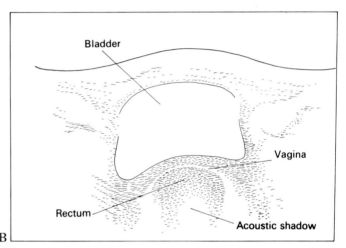

Fig. 10.38 Ballooned rectum. Rectal gas posterior to the vagina.
A. Sagittal section
B. Transverse section.

THE PERITONEAL RECESSES

The peritoneal cavity extends from the diaphragm above to the pelvic floor below. It is bounded anteriorly and laterally by the abdominal wall and posteriorly by the retroperitoneal organs. It is divided into the greater sac, or general peritoneal cavity, and the lesser sac which lies behind the stomach and the gastrohepatic and gastrocolic ligaments. The general peritoneal cavity is divided into supramesocolic and inframesocolic compartments by the transverse mesocolon. The inframesocolic space is further divided by the root of the mesentery into a smaller right and a larger left infracolic space, the latter communicating with the pelvic cavity. In these compartments there are recesses formed by the peritoneal reflections over abdominal organs. These compartments can only be recognised ultrasonically when they are distended with fluid.

The *transverse mesocolon* ascends from the colon to be attached to the descending part of the duodenum, the head and lower margins of the body of the pancreas and the anterior surface of the left kidney.

The *root of the mesentery* is short (15 cm) and passes obliquely across the posterior abdominal wall from the duodenojejunal junction to the ileocaecal junction. It crosses the anterior surface of the horizontal part of the duodenum, the aorta, the inferior vena cava, the right ureter and right psoas muscle.

In the supracolic compartment there are seven important recesses recognised around the liver. The *right subhepatic space* or *right posterior intraperitoneal space* lies transversely behind the right lobe of the liver. It is divided into an anterior and posterior compartment (Morison's pouch) and is bounded on the right by the right lobe of the liver and the diaphragm. To the left is the epiploic foramen of Winslow and inferiorly the descending part of the duodenum. Anteriorly it is bounded by the liver and gall bladder and posteriorly by the upper part of the right kidney and diaphragm. Superiorly the space is bounded by the liver, inferiorly by the transverse colon.

The *right anterior intraperitoneal space* or *right subphrenic space* lies between the liver and the diaphragm; it is bounded posteriorly by the coronary ligament and right lateral ligament and to the left by the falciform ligament.

The *right extraperitoneal space* is the bare area of the liver (see page 63).

The *left subphrenic space* or *left anterior intraperitoneal space* surrounds the left lobe of the liver and has freely communicating subphrenic and subhepatic components. It is bounded above by the diaphragm and behind by the left lateral ligament and the left lobe of the liver, the gastrohepatic omentum and the anterior surface of the stomach. To the right is the falciform ligament, to the left is the spleen, gastrosplenic omentum and the diaphragm.

The *left posterior intraperitoneal space (the lesser sac)* is entered through the epiploic foramen of Winslow. The anterior margin of the foramen is formed by the right border of the gastrohepatic omentum, which contains the portal vein, common bile duct and hepatic artery. The inferior vena cava lies posteriorly. The epiploic foramen lies above the body of the pancreas approximately at the level of the coeliac axis. The *lesser sac* is bounded behind by the diaphragm, the pancreas, the transverse mesocolon and transverse colon. In front lies the liver, the gastrohepatic omentum and stomach and to the left is the lienorenal ligament, the spleen and the gastrosplenic omentum. The superior recess of the lesser sac which lies behind the quadrate lobe of the liver is divided from the larger recess, the body, by a fold of peritoneum containing the left gastric artery. The inferior recess of the body extends for a variable distance into the greater omentum.

Below the mesocolon the compartments are divided into paracolic, infracolic and pelvic spaces. The *paracolic gutters*, which are the lateral part of the paravertebral gutters, form a route of communication between the supramesocolic and inframesocolic compartments and are continuous with the pelvic cavity. The *right paracolic gutter* is wider than the left and communicates above with the right subhepatic space, particularly its posterior extension. The left paracolic gutter is limited above by the phrenicocolic ligament, extending from the splenic flexure of the colon to the diaphragm, which separates it from the left subphrenic space.

The pelvis is the most dependent part of the peritoneal cavity. The posterior surface is covered with peritoneum down to the level of the second sacral vertebra where it is reflected forward off the rectum onto the pelvic viscera forming pouches. In the female the *rectouterine pouch of Douglas* (the cul-de-sac) lies behind the uterus and upper third of the vagina, and anterior to the uterus is the *uterovesical pouch*. In the male there is one pouch, the *rectovesical pouch*. Anteriorly in both male and female *paravesical fossae* lie on either side of the bladder formed by reflections of the peritoneum to the side wall of the pelvis. The pelvic cavity is further divided by peritoneal reflections to the side wall of the pelvis, the rectouterine and broad ligaments in the female and rectovesical fold in the male.

Though these recesses are not seen ultrasonically in a normal abdomen a knowledge of their location is essential, as they provide watersheds and drainage basins for the spread and localisation of infection (Meyers 1976).

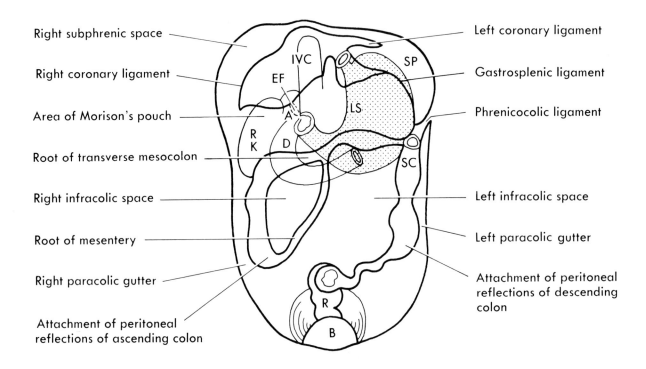

Fig. 10.39   The posterior peritoneal reflections and recesses.
IVC —Inferior vena cava   D   —Duodenum
EF —Epiploic foramen of Winslow LS —Lesser sac
RK —Right kidney   SP —Spleen
A   —Right adrenal   SC —Splenic flexure of colon
The position of the stomach is shown as a dotted area.

THE GASTRO-INTESTINAL TRACT AND PERITONEAL CAVITY 165

In the following illustrations the peritoneal recesses are indicated by arrows.

THE RIGHT SUBHEPATIC SPACE OR RIGHT POSTERIOR INTRAPERITONEAL SPACE

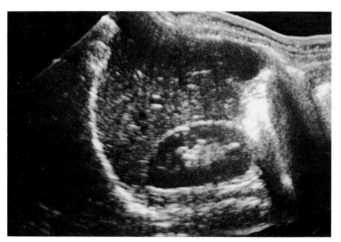 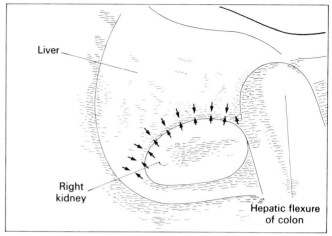

Fig. 10.40A The right subhepatic spaces or right posterior intraperitoneal space. Sagittal section: both the anterior and posterior compartments of the subhepatic space are defined on a sagittal section.

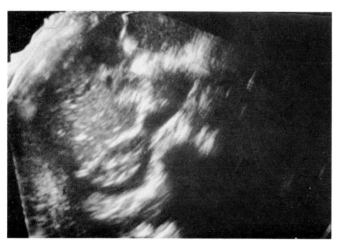 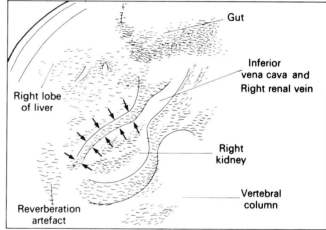

Fig. 10.40B The right subhepatic spaces or right posterior intraperitoneal space. Low transverse section: the anterior compartment or anterior subhepatic space.

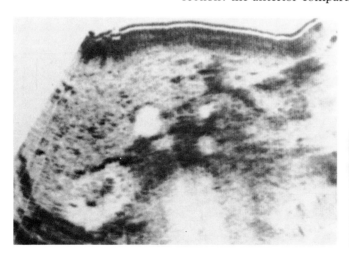 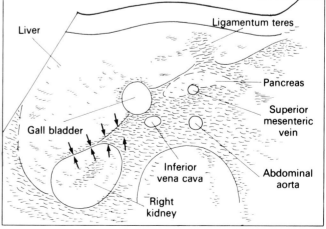

Fig. 10.40C The right subhepatic spaces or right posterior intraperitoneal space. Mid-transverse section: the anterior compartment or anterior subhepatic space.

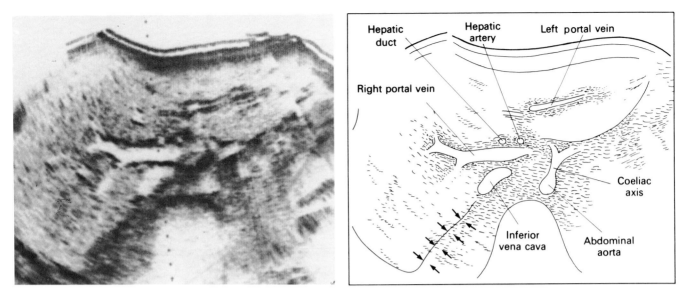

Fig. 10.40D The right subhepatic spaces or right posterior intraperitoneal space. High transverse section: the posterior subhepatic space (Morison's pouch).

RIGHT SUBPHRENIC SPACE OR RIGHT ANTERIOR INTRAPERITONEAL SPACE

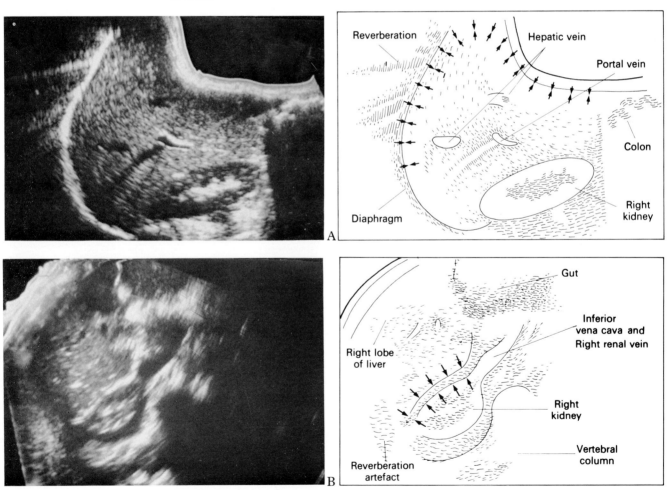

Fig. 10.41 The right subphrenic space or right anterior intraperitoneal space. This is bounded posteriorly by the coronary ligament and right lateral ligament. It is separated from the left subphrenic space by the falciform ligament, but the two spaces communicate below its free margin.
  A. Sagittal section: right subphrenic space
  B. Transverse section: right and left subphrenic spaces

THE GASTRO-INTESTINAL TRACT AND PERITONEAL CAVITY 167

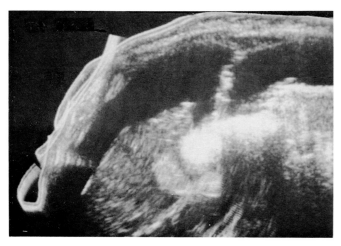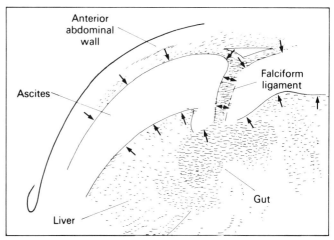

Fig. 10.42 The falciform ligament. This separates the right and left subphrenic spaces. In this subject ascites is present and the subphrenic spaces are displayed.

THE LEFT SUBPHRENIC SPACE OR LEFT ANTERIOR INTRAPERITONEAL SPACE

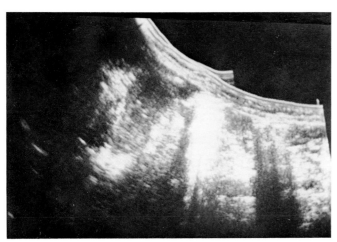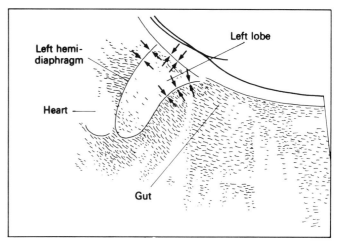

Fig. 10.43A Sagittal section. The left anterior intraperitoneal space or left subphrenic space is bounded superiorly by the coronary ligament and inferiorly by the lesser omentum. It has two freely communicating anterior and posterior compartments. It is separated from the right subphrenic space by the falciform ligament.

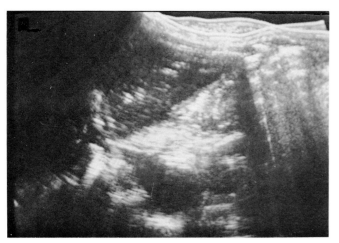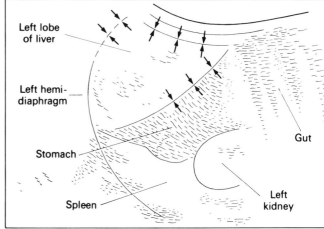

Fig. 10.43B Sagittal section. (Refer Fig. 10.43A caption.)

168 ULTRASONIC SECTIONAL ANATOMY

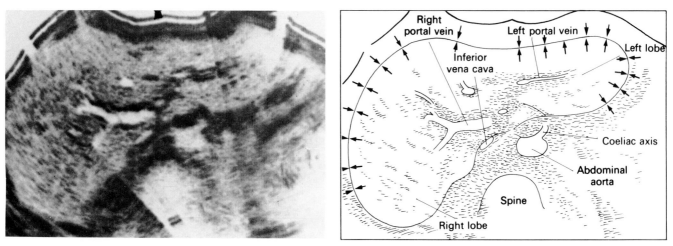

Fig. 10.43C High transverse section with right and left subphrenic spaces. (Refer Fig. 10.43A caption.)

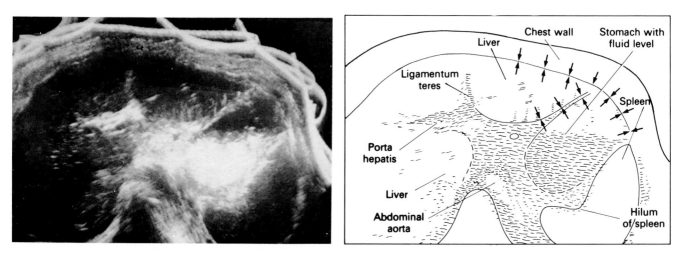

Fig. 10.43D Low transverse section. (Refer Fig. 10.43A caption.)

THE EPIPLOIC FORAMEN OF WINSLOW

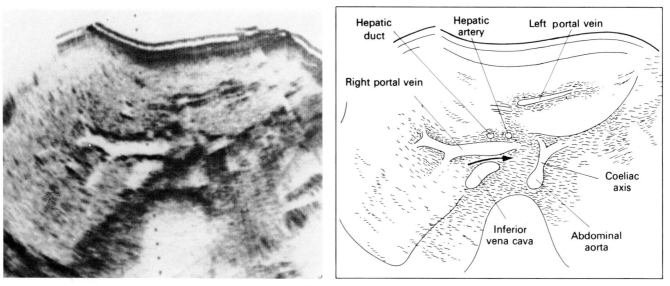

Fig. 10.44A The epiploic foramen of Winslow or aditus to the lesser sac lies approximately at the level of the coeliac axis. The right and left posterior intraperitoneal spaces (Morison's pouch and the lesser sac) communicate through the foramen. Transverse section.

THE GASTRO-INTESTINAL TRACT AND PERITONEAL CAVITY   169

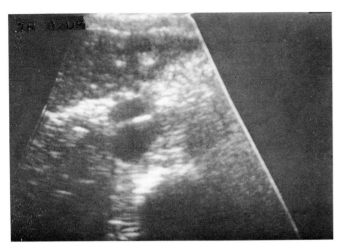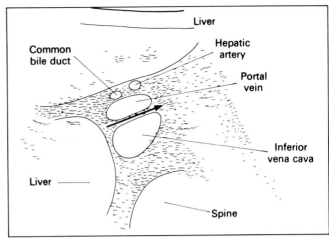

Fig. 10.44B The epiploic foramen of Winslow or aditus to the lesser sac lies approximately at the level of the coeliac axis. The right and left posterior intraperitoneal spaces (Morison's pouch and the lesser sac) communicate through the foramen. Transverse section.

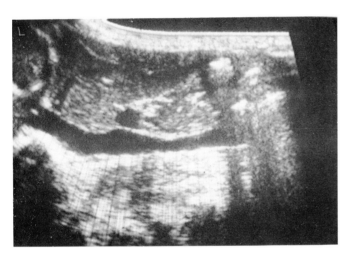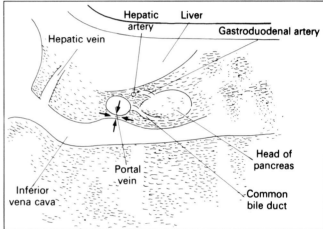

Fig. 10.44C The epiploic foramen of Winslow or aditus to the lesser sac lies approximately at the level of the coeliac axis. The right and left posterior intraperitoneal spaces (Morison's pouch and the lesser sac) communicate through the foramen. Sagittal section.

THE LESSER SAC OR LEFT POSTERIOR INTRAPERITONEAL SPACE

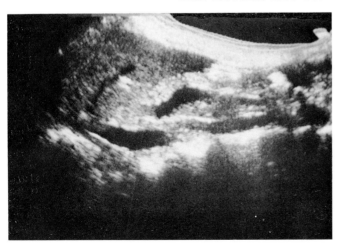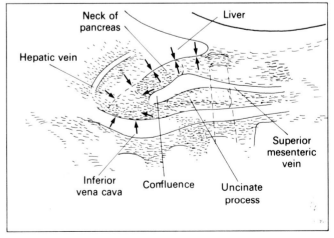

Fig. 10.45A The lesser sac or left posterior intraperitoneal space. Sagittal section through the superior recess.

170 ULTRASONIC SECTIONAL ANATOMY

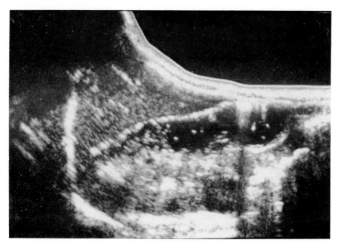 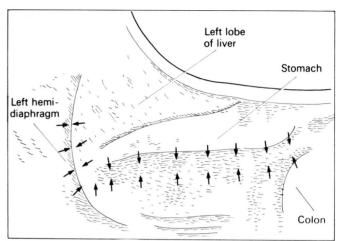

Fig. 10.45B The lesser sac or left posterior intraperitoneal space. Sagittal section through the body.

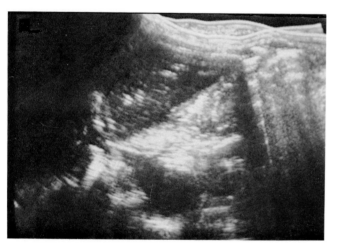 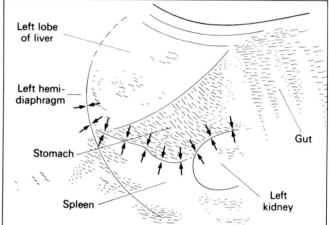

Fig. 10.45C The lesser sac or left posterior intraperitoneal space. Sagittal section through the body.

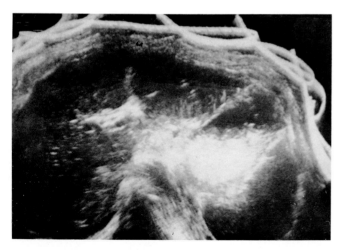 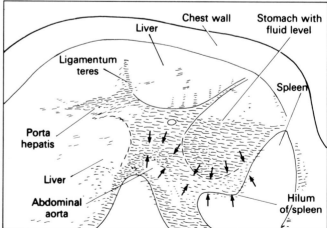

Fig. 10.45D The lesser sac or left posterior intraperitoneal space. Transverse section at the level of the body.

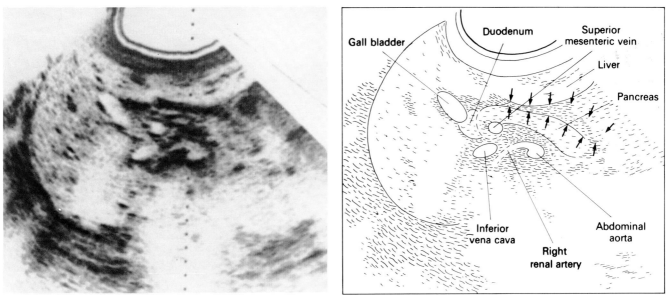

Fig. 10.45E The lesser sac or left posterior intraperitoneal space. Transverse section at the level of the inferior recess.

THE PARACOLIC GUTTERS

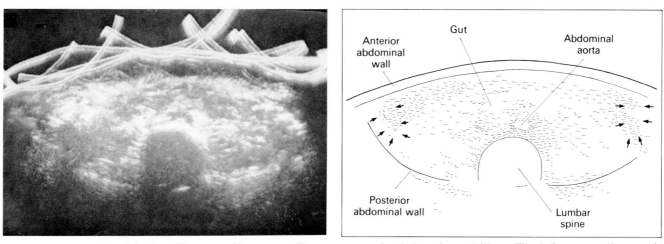

Fig. 10.46 The paracolic gutters. Transverse section below the umbilicus. The inframesocolic space is anterior to the aorta. (See Fig. 10.1.)

THE RECTOUTERINE POUCH OF DOUGLAS

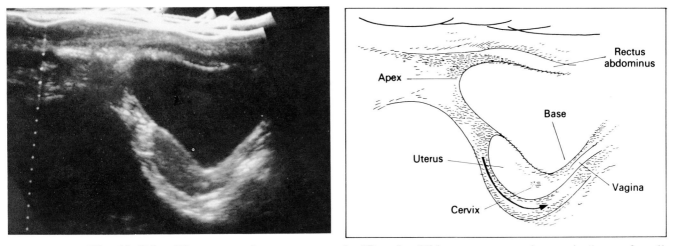

Fig. 10.47A The rectouterine space or pouch of Douglas. This recess commonly contains loops of small bowel. Longitudinal section.

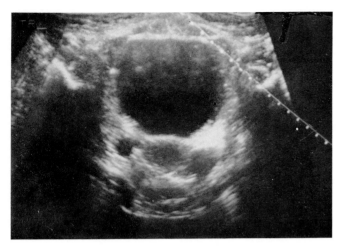 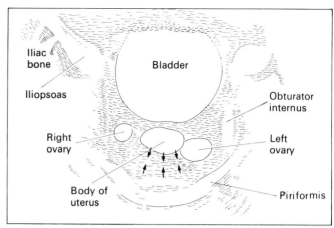

Fig. 10.47B The rectouterine space or pouch of Douglas. This recess commonly contains loops of small bowel. Transverse section.

THE UTEROVESICAL POUCH

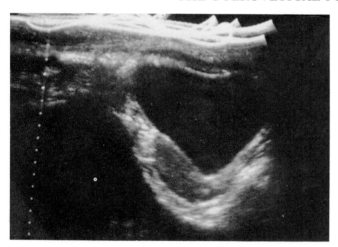 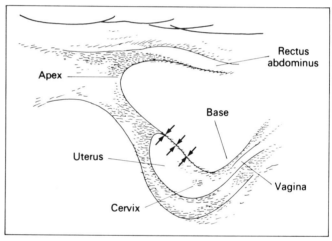

Fig. 10.48A The uterovesical pouch. This recess is normally empty. When the uterus is retroverted it may contain loops of small intestine. Longitudinal section.

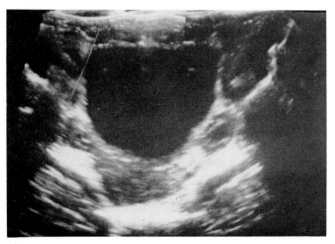 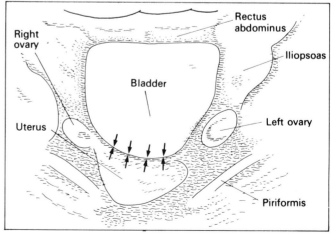

Fig. 10.48B The uterovesical pouch. This recess is normally empty. When the uterus is retroverted it may contain loops of small intestine. Transverse section.

THE GASTRO-INTESTINAL TRACT AND PERITONEAL CAVITY 173

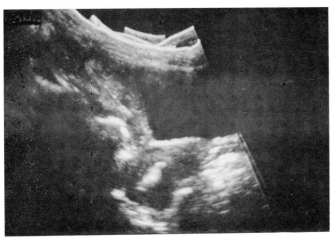 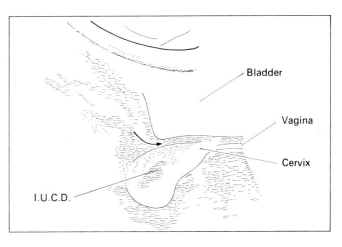

Fig. 10.48C The uterovesical pouch. This recess is normally empty. When the uterus is retroverted it may contain loops of small intestine. Longitudinal section with the uterus retroverted.

THE RECTOVESICAL POUCH

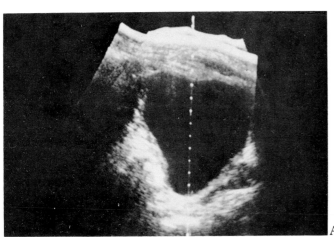 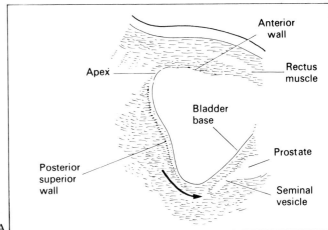

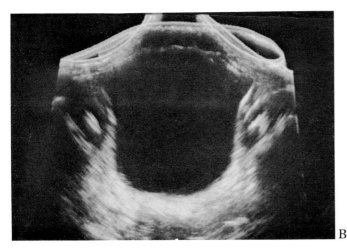 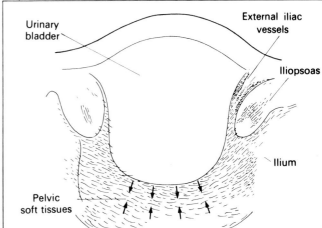

Fig. 10.49 The rectovesical pouch.
A. Longitudinal section
B. Transverse section.

11 THE MAJOR VESSELS

THE MAJOR VISCERAL ARTERIES AND VEINS

The abdominal aorta and major branches

The *abdominal aorta* descends on the vertebral column from the aortic hiatus in the diaphragm to its bifurcation at the level of the fourth lumbar vertebra. It lies slightly to the left of the vertebral column and in slim subjects the inferior part is very close to the abdominal wall. The upper width of normal for the aorta is 3.0 cm at the level of the diaphragm and 2.5 cm at the umbilicus. The aorta is almost invariably seen on ultrasonic sections in the upper third of its abdominal course but lower down it may be obscured by intestinal gas. It usually lies within 1 cm of the anterior border of the vertebral column, and a greater distance may be of pathological significance. Several of its major branches can be seen arising from its anterior and lateral aspect, and these include the coeliac axis, the superior mesenteric artery, the renal arteries and occasionally the inferior mesenteric artery. The iliac arteries can also be seen after the bifurcation but for a short distance only.

The *coeliac axis or trunk* is the most cephalic of the unpaired branches of the aorta. It arises from the aorta at an angle of about 60° and runs forward for about 1 cm before dividing into three vessels, the common hepatic, the splenic and the left gastric arteries. These supply the gut from the lower oesophagus to the descending duodenum and the liver.

The *superior mesenteric artery* is the second median branch of the aorta supplying the gut. It arises 0.5 cm inferior to the coeliac axis at the level of the first lumbar vertebra. It usually runs parallel with and anterior to the aorta behind the body of the pancreas and descends anterior to the left renal vein and the horizontal part of the duodenum and runs obliquely in the root of the mesentery to the right iliac fossa. This artery supplies the gut from the descending duodenum to the transverse colon.

The *renal arteries* originate approximately at the level of the upper part of the second lumbar vertebra in the region of the pancreas. They cross the corresponding crus of the diaphragm and psoas muscle, to supply the kidney and send small branches to the ureter and suprarenal gland. The right renal artery passes behind the inferior vena cava.

The *inferior mesenteric* artery is the third and last median branch of the aorta to the gut. It arises posterior to the horizontal part of the duodenum and descends on the left side of the aorta posterior to the peritoneum. It supplies the intestine from the transverse colon to the anal canal.

The *common iliac arteries* are the terminal branches of the abdominal aorta. They arise at the level of the fourth lumbar vertebra and pass inferolaterally to the superior surface of the sacroiliac joint and divide into internal and external iliac arteries. The right common iliac artery lies directly on the right common iliac vein and the beginning of the inferior vena cava.

The *external iliac arteries* are a direct continuation of the common iliac arteries. They pass inferolaterally along the superior aperture of the lesser pelvis, medial to and then anterior to psoas, in the mid-inguinal point and pass posterior to the inguinal ligament to become the femoral artery.

The *internal iliac artery* is the smaller of the two branches of the common iliac artery in the adult. It is larger in the fetus as it transmits blood to the placenta through the umbilical artery. In the adult it supplies some of the contents of the lesser pelvis, the perineum, the greater part of the gluteal region and the iliac fossa. It passes posteriorly into the pelvis medial to the external iliac vein, lies between the ureter inferiorly and the internal iliac vein superiorly, and divides into numerous posterior and anterior branches.

The inferior vena cava and tributaries

The *inferior vena cava* is the widest vein in the body. It drains blood from the lower limbs, the urogenital systems, the suprarenal glands and the liver.

It begins at the level of the fifth lumbar vertebra behind the right common iliac artery. It is formed by the union of the two common iliac veins. It ascends on the right side of the median plane anterior to the vertebral column and right psoas muscle. Superiorly it curves forward on the right crus of the diaphragm, to reach a deep groove on the posterior aspect of the liver between the right lobe and the caudate lobe. Posteriorly it is related to the sympathetic trunk, the lumbar arteries, the right renal artery, the coeliac ganglion, the middle suprarenal artery and the right suprarenal gland and the diaphragm. As it ascends it lies behind the root of the mesentery, the ileocolic and right colic vessels, the duodenum, the head of the pancreas and bile duct, the portal vein, the superior part of the duodenum and the liver. The main tributaries are the common iliac veins, the renal veins and the hepatic veins.

Providing a longitudinal section is taken in full inspiration, preferably using the Valsalva manoeuvre, the inferior vena cava is seen on ultrasonic sections posterior to the liver. As with the aorta it frequently cannot be seen at a lower level due to intestinal gas. It takes a curving course through the liver and lies immediately adjacent and posterior to the portal vein. It may show pulsation due to transmitted cardiac pressure changes. No satisfactory upper limits of normal size have been defined for the inferior vena cava and its branches, but absence of the normal size changes with respiration is suggestive of congestive heart failure. Above the level of the renal veins the size of the inferior vena cava is increased in normal subjects. The only major tributaries are the *hepatic veins* and *renal veins*. On longitudinal section the middle hepatic vein takes a longitudinal course through the middle of the liver—on transverse sections the right and left hepatic veins take a transverse course from the lateral portions of the liver. The *renal veins* are seen at the same level as the renal arteries, towards the inferior aspect of the pancreas. The right renal vein has a short course directly to the right kidney. The left renal vein passes between the superior mesenteric artery and the abdominal aorta on a long course to the left kidney. There is often a noticeable difference in size of the left renal vein before it passes through the narrow area between the superior mesenteric artery and aorta and a pressure difference along the renal vein can be measured at venography in many individuals.

The extrahepatic portal system

The *portal vein* begins posterior to the junction of the body and the head of the pancreas. It is formed by the union of the splenic and superior mesenteric veins. It ascends behind the superior part of the duodenum receiving the pancreaticoduodenal and the right and left gastric veins. It ascends in the free edge of the lesser omentum posterior to the common bile duct and the hepatic artery, and divides into right and left branches at the porta hepatis. The right branch forms the continuation of the main portal vein, receives the cystic vein and runs horizontally into the liver. The left portal vein is angled acutely passing anteriorly towards the left uniting with the ligamentum teres and ligamentum venosum before entering the left lobe of the liver.

The *splenic vein* is formed from five or six tributaries which leave the hilus of the spleen. It passes to the right on the posterior surface of the pancreas lying on the left kidney, the left psoas muscle, the left crus of the diaphragm and the abdominal aorta, between the origins of the coeliac trunk and superior mesenteric artery. It joins the superior mesenteric vein to form the portal vein anterior to the inferior vena cava. The splenic vein is an important ultrasonic landmark for the identification of the pancreas.

The *superior mesenteric vein* lies to the right of the superior mesenteric artery in the root of the mesentery. It drains blood from the territory of the artery and receives the right gastro-epiploic vein. Sometimes the inferior mesenteric vein and pancreaticoduodenal vein join the superior mesenteric vein. Superiorly the superior mesenteric vein passes to the right, anterior to the uncinate process of the pancreas and behind the neck of the pancreas, to join the splenic vein and form the main portal vein. The superior mesenteric vein, like the splenic vein, is an important ultrasonic landmark used in localisation of the pancreas.

The *inferior mesenteric vein* ascends from the pelvis lateral to the inferior mesenteric artery. It lies lateral to the duodenojejunal flexure and anterior to the left renal vein. It joins either the splenic vein behind the pancreas or may curve medially and to the right to join the superior mesenteric vein or the confluence of the splenic and superior mesenteric veins.

THE ABDOMINAL AORTA AND MAJOR BRANCHES

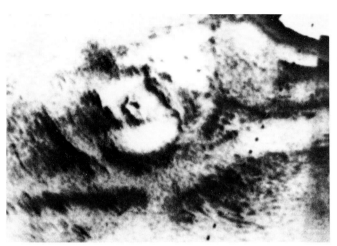 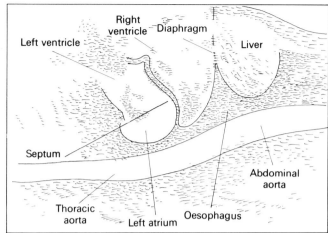

Fig. 11.1 Oblique paramedian section. The thoracic aorta lies posterior to the pericardium to the left side of the posterior mediastinum. It inclines anteriorly and to the right behind the oesophagus to enter the abdomen posterior to the median arcuate ligament of the diaphragm at the level of the 12th thoracic vertebra.

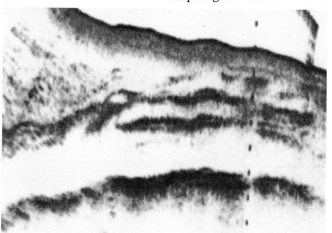 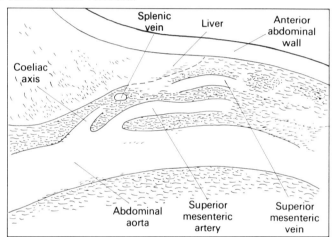

Fig. 11.2A Sagittal section. The coeliac axis and superior mesenteric artery are proximal unpaired ventral branches of the abdominal aorta which originate just below the diaphragm. The splenic vein passes anterior to the aorta between the two arteries.

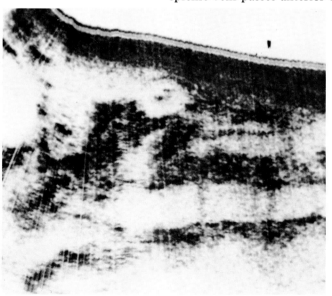 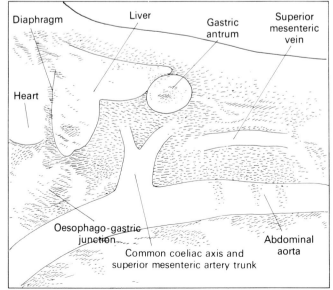

Fig. 11.2B Sagittal section. The coeliac axis and superior mesenteric artery have a common origin from the abdominal aorta.

178 ULTRASONIC SECTIONAL ANATOMY

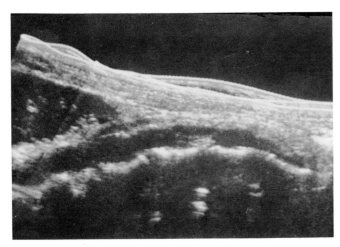 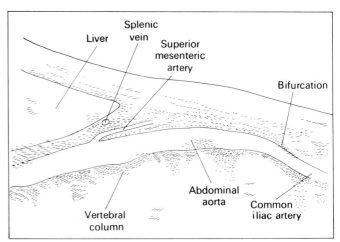

Fig. 11.3 Sagittal section. The abdominal aorta is outlined from just below the diaphragm to its bifurcation into the common iliac arteries at the level of the fourth lumbar vertebra. A short section of the common iliac artery is outlined below the bifurcation.

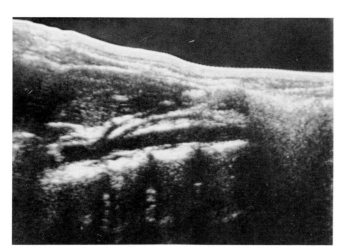 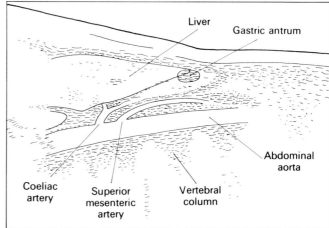

Fig. 11.4A Sagittal section. Normal aorta. The abdominal aorta has a smooth outline and uniform calibre.

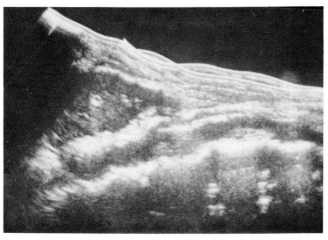 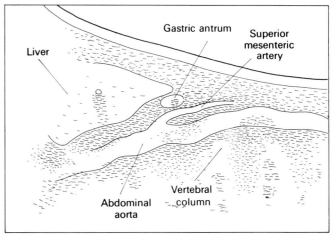

Fig. 11.4B Sagittal section. Atheromatous aorta. The outline of the aorta is slightly irregular with variations in calibre.

THE MAJOR VESSELS 179

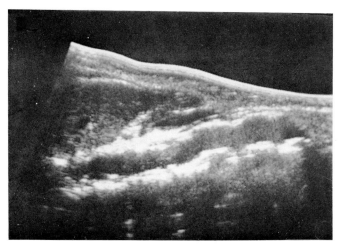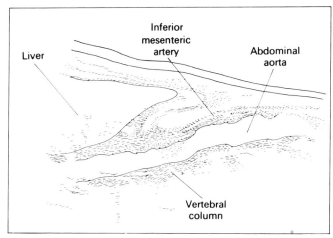

Fig. 11.4C Sagittal section. Atheromatous dilated aorta. The aorta is atheromatous and its calibre increases inferiorly. The origin of the inferior mesenteric artery is seen on this section.

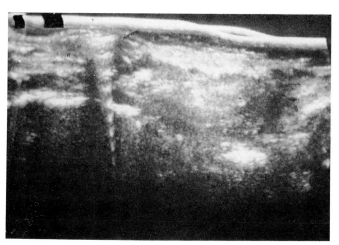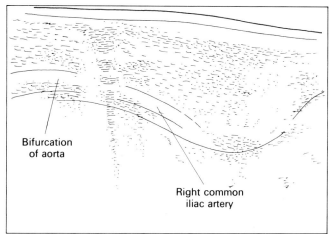

Fig. 11.5 Oblique section from the bifurcation of the aorta towards the mid-inguinal point. The common iliac artery is partially outlined, below this level detail is obscured by overlying intestinal contents.

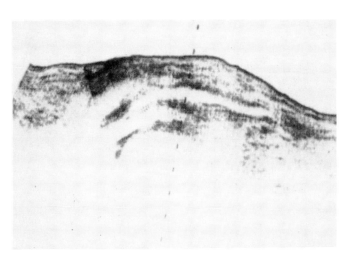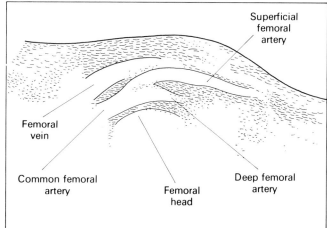

Fig. 11.6 Oblique section below the inguinal ligament. The femoral vessels are outlined anterior to the head of the femur. They are superficial and are usually easily identified.

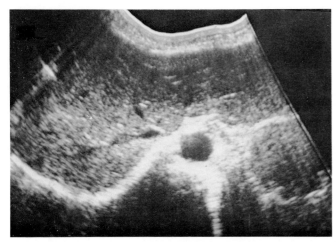 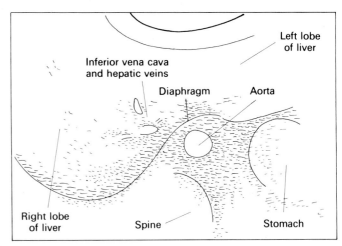

Fig. 11.7 High transverse section. The abdominal aorta is seen posteriorly behind the median arcuate ligament of the diaphragm and anterior to the spine. This section is at the level of the 12th dorsal vertebra.

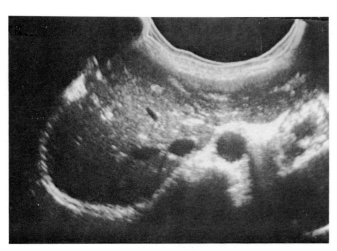 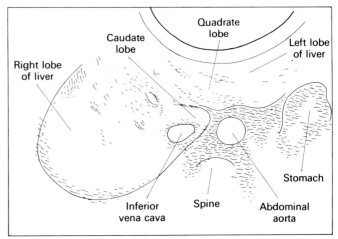

Fig. 11.8 High transverse section. The abdominal aorta lies to the left of the inferior vena cava behind the left lobe of the liver.

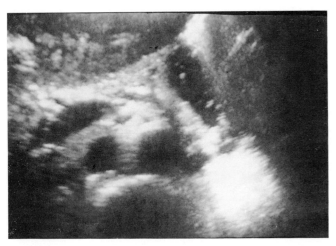 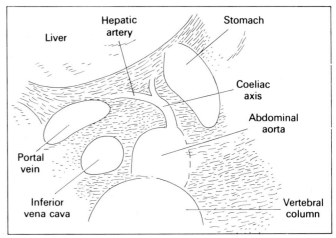

Fig. 11.9 Transverse section. The origin of the coeliac axis has a Y-shaped configuration in the plane. The common hepatic artery passes to the right anterior to the portal vein, the left arm of the Y is the origin of the splenic artery.

THE MAJOR VESSELS 181

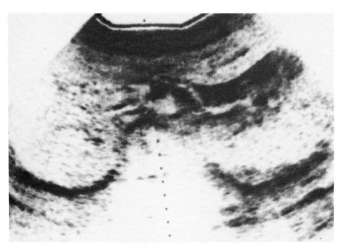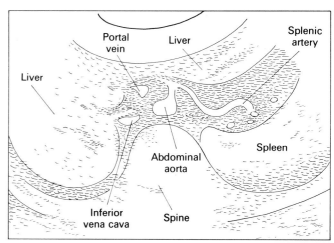

Fig. 11.10 Transverse section. The splenic artery runs to the left towards the hilum of the spleen where it divides into five or six branches. It is the largest branch of the coeliac axis and lies just above the superior border of the pancreas. The additus and body of the lesser sac lie in this plane.

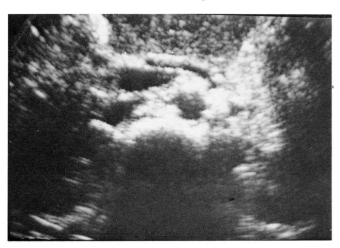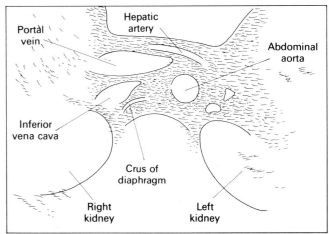

Fig. 11.11 Transverse section. The common hepatic artery is anterior to the portal vein. At this point it divides into the hepatic artery and gastroduodenal artery and then ascends through the lesser omentum to the porta hepatis lying anterior to the portal vein and to the left of the bile duct. (See Figs. 7.65 and 7.66.)

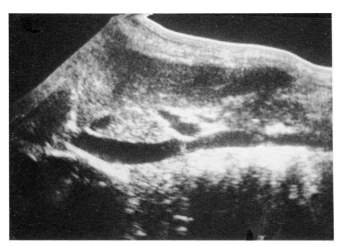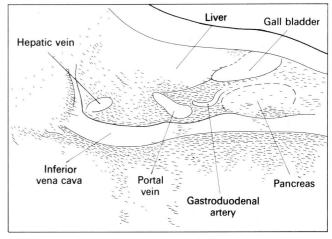

Fig. 11.12 Sagittal section. The gastroduodenal artery is a branch of the common hepatic artery. It descends posterior to the superior duodenum and anterior to the pancreas. It is a useful anatomic landmark for identifying the head of the pancreas. (See Fig. 7.67.)

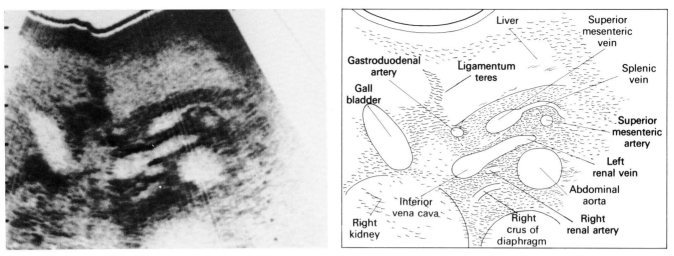

Fig. 11.13 Transverse section between the first and second lumbar vertebrae. The abdominal aorta lies posteriorly behind the pancreas, the splenic vein, superior mesenteric vessels and the left renal vein. Note the gastroduodenal artery which lies anterolateral to the head of the pancreas.

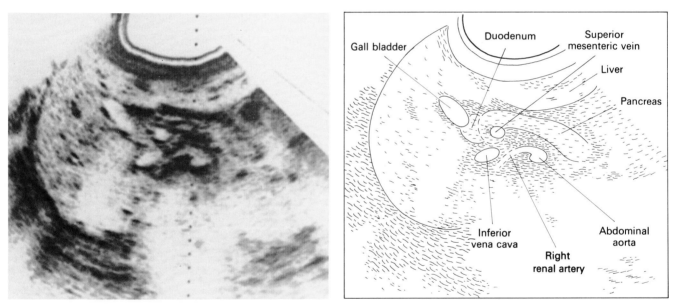

Fig. 11.14 Transverse section. The origin of the right renal artery is clearly demonstrated behind the pancreas. The artery passes to the right kidney behind the inferior vena cava. This section is approximately at the superior border of the second lumbar vertebra. Only the origin of the left renal artery is seen, distally it is obscured by the renal vein. See Figure 11.34.

THE MAJOR VESSELS 183

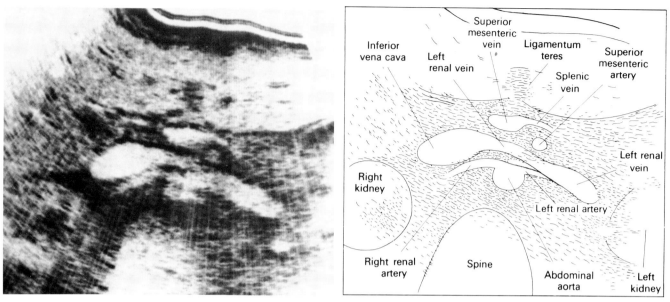

Fig. 11.15 Transverse section at the origin of both renal arteries. Note that the right renal artery passes behind the inferior vena cava and that the left renal vein is behind the superior mesenteric artery.

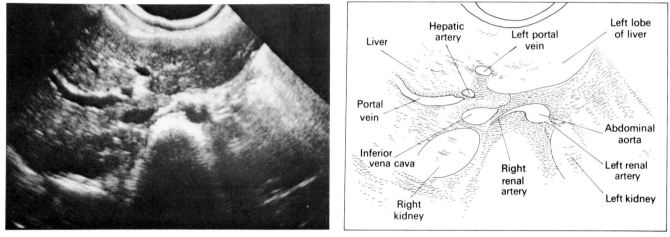

Fig. 11.16 Transverse section. The right renal artery passes behind the inferior vena cava to supply the right kidney.

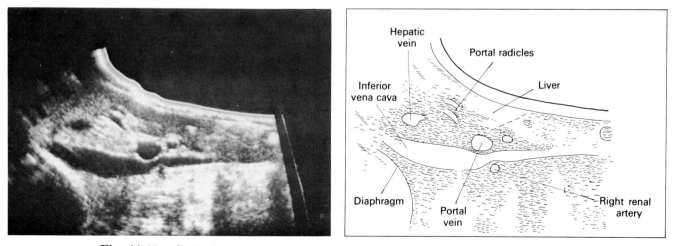

Fig. 11.17 Sagittal section. The right renal artery is behind the inferior vena cava.

184 ULTRASONIC SECTIONAL ANATOMY

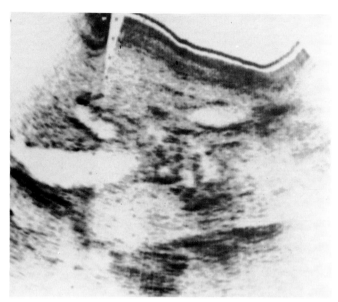 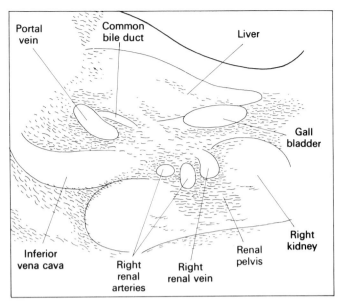

Fig. 11.18 Oblique sagittal section. Two right renal arteries and the renal vein lie in the hilum of the right kidney, lateral to the inferior vena cava.

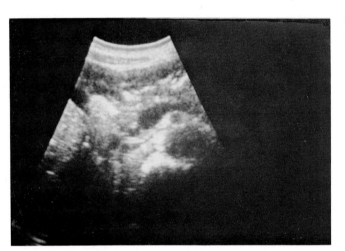 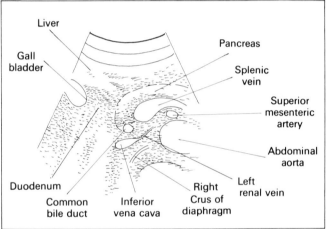

Fig. 11.19 Limited transverse section. The superior mesenteric artery lies anterior to the aorta behind the splenic vein and the pancreas.

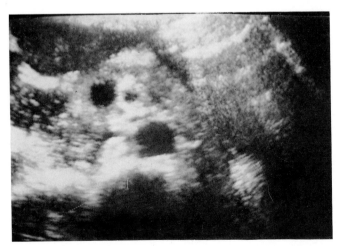 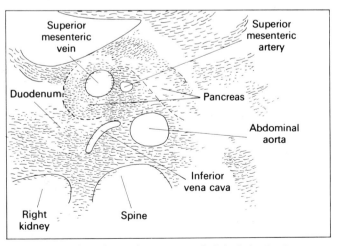

Fig. 11.20 Transverse section. The superior mesenteric vein and artery are behind the body of the pancreas anterior to the aorta.

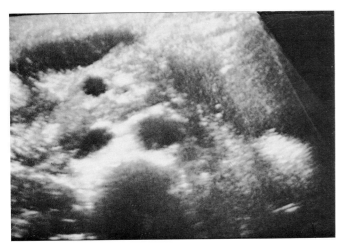 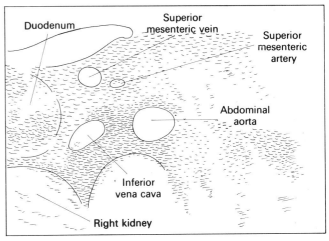

Fig. 11.21 Transverse section. The superior mesenteric artery and vein are in the root of the mesentery adjacent to the descending duodenum. They are not defined below this level.

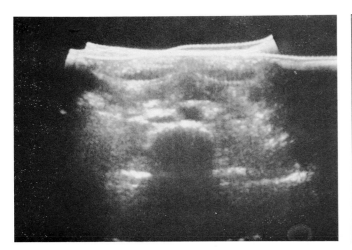 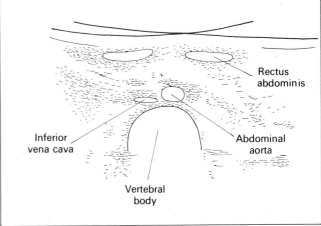

Fig. 11.22 Transverse section at the level of the umbilicus above the bifurcation of the aorta.

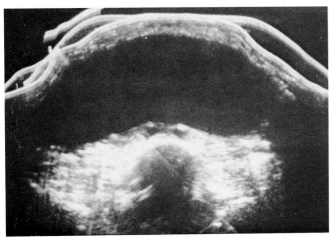 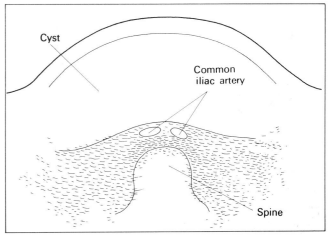

Fig. 11.23 Transverse section just below the bifurcation of the abdominal aorta into right and left common iliac arteries. This patient had a cyst in the lower abdomen and as a result the bifurcation is clearly demonstrated. It is commonly obscured by intestinal contents.

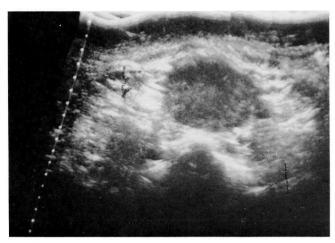 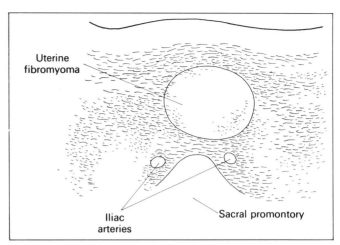

Fig. 11.24 Transverse section at the level of the sacral promontory. The common iliac vessels are outlined posteriorly. In this section a uterine fibroid is displacing the small intestine and acting as an acoustic window.

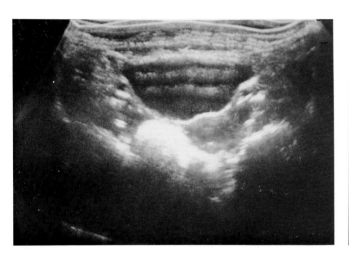 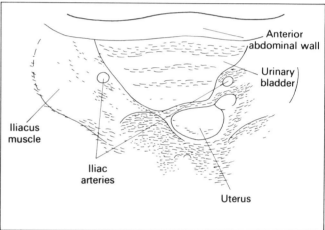

Fig. 11.25 Transverse section. The common iliac arteries lie medial to the iliopsoas muscle. This section is approximately at the level of their division into internal and external iliac arteries.

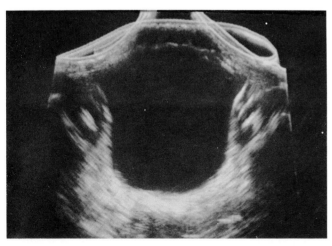 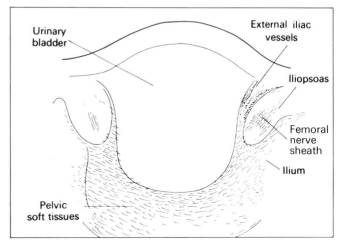

Fig. 11.26 Transverse section through the lesser pelvis. The external iliac vessels are medial to the psoas muscle.

THE MAJOR VESSELS 187

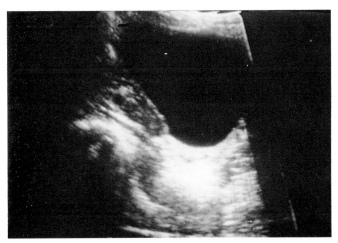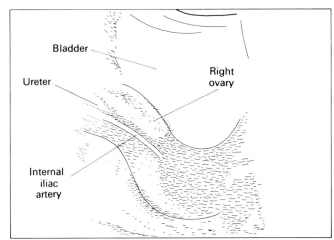

Fig. 11.27 Sagittal section. The internal iliac vessels are posterior to the ovary. The ureter crosses the vessels above the ovary. (See also Fig. 9.47B.)

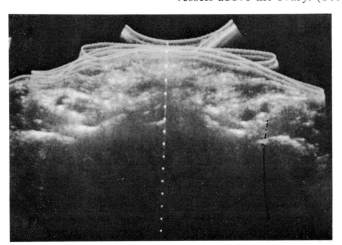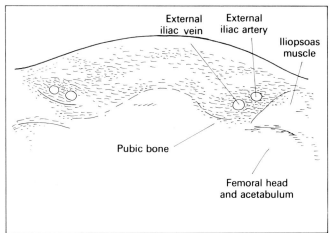

Fig. 11.28 Transverse section above the inguinal ligament. The vessels are more superficial and the iliac artery and vein are outlined medial to iliopsoas.

THE INFERIOR VENA CAVA AND MAJOR TRIBUTARIES

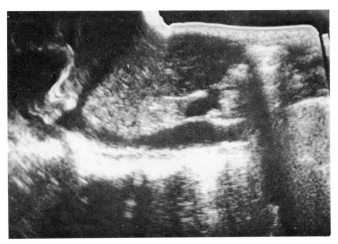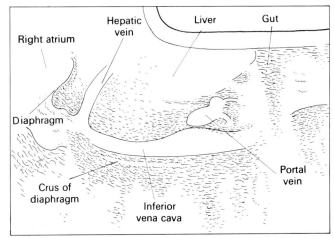

Fig. 11.29 Sagittal section in the upper abdomen. The upper part of the inferior vena cava is outlined behind the liver. Below the liver it is obscured by intestinal gas. The middle hepatic vein enters the inferior vena cava just below the central tendon of the diaphragm. The inferior vena cava pierces the diaphragm at the level of the eighth dorsal vertebra to enter the right atrium.

188 ULTRASONIC SECTIONAL ANATOMY

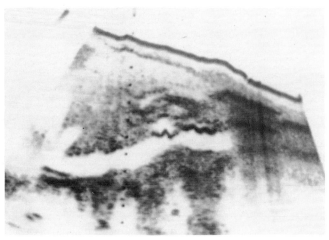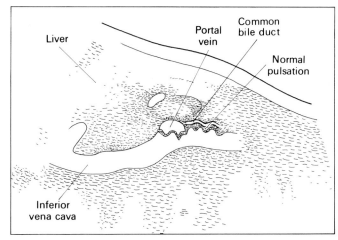

Fig. 11.30 Sagittal section. Normal pulsation from cardiac pressure changes have been recorded with a slow linear scan.

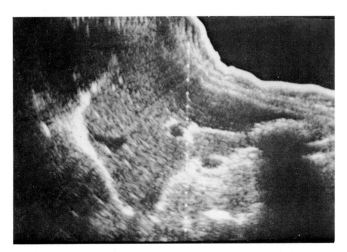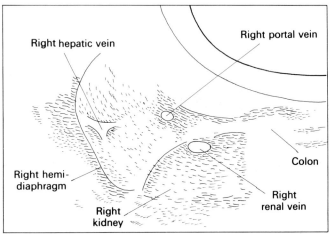

Fig. 11.31 Sagittal section to the right of the inferior vena cava. The right renal vein is leaving the hilum of the right kidney.

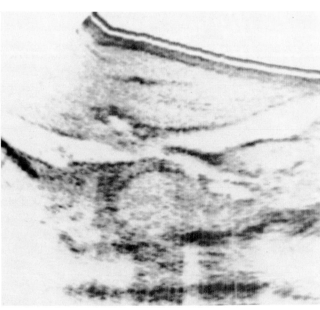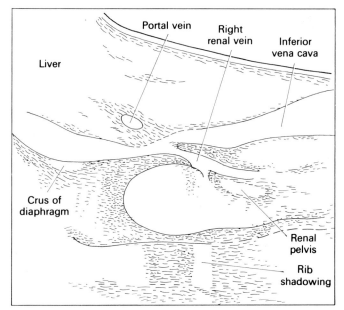

Fig. 11.32 Sagittal section with lateral angulation in the plane of the inferior vena cava. The right renal vein enters the inferior vena cava inferior to the level of the portal vein.

THE MAJOR VESSELS 189

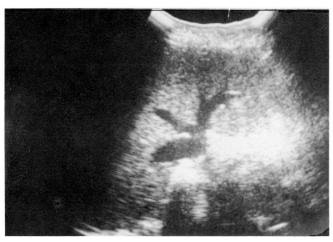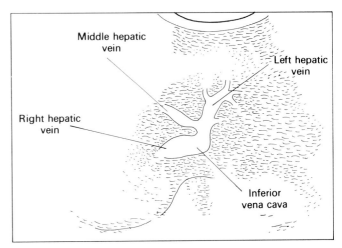

Fig. 11.33 Transverse section in the epigastrium. The middle and left hepatic veins commonly join before entering the inferior vena cava. There are usually three main hepatic veins and several smaller veins draining into the inferior vena cava.

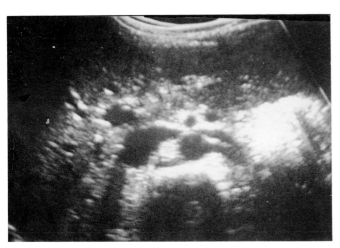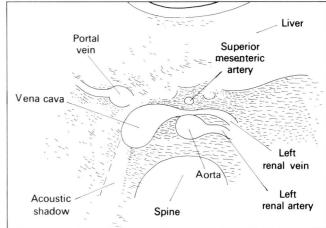

Fig. 11.34 Transverse section. The left renal vein passes anterior to the aorta and behind the superior mesenteric artery. It lies anterior to the left renal artery.

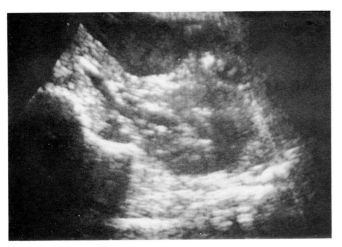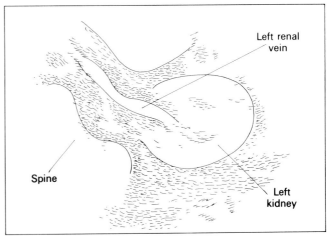

Fig. 11.35 Transverse section. The left renal vein is longer than the right renal vein. In this section it passes medially from the hilum of the kidney anterior to the spine to pass anterior to the abdominal aorta before entering the vena cava.

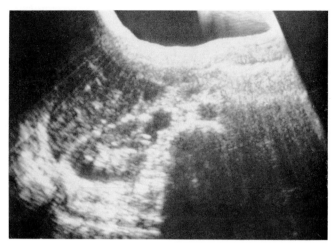 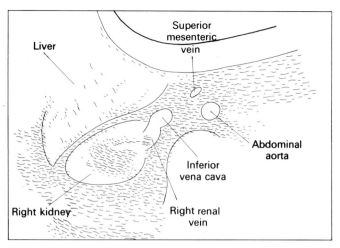

Fig. 11.36 Transverse section. The right renal vein leaves the right renal hilum and enters the vena cava posterolaterally after a short course.

THE EXTRAHEPATIC PORTAL SYSTEM

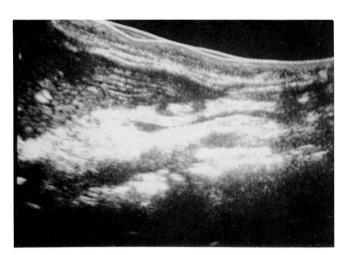 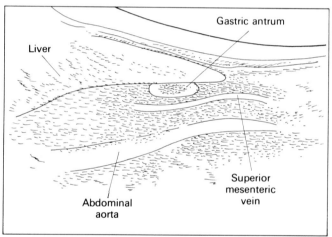

Fig. 11.37 Sagittal section. The superior mesenteric vein passes behind the stomach lying in the mesentery anterior to the abdominal aorta. Superiorly it curves to the right towards the inferior vena cava.

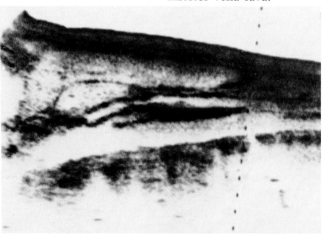 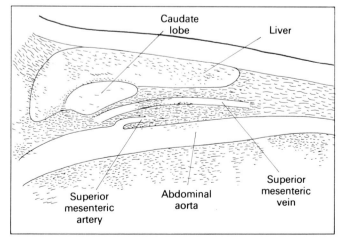

Fig. 11.38 Sagittal section. The superior mesenteric vein lies anterior to the superior mesenteric artery in front of the aorta. It passes above the origin of the artery and curves posteriorly to join the splenic vein behind the body of the pancreas to form the portal vein.

THE MAJOR VESSELS 191

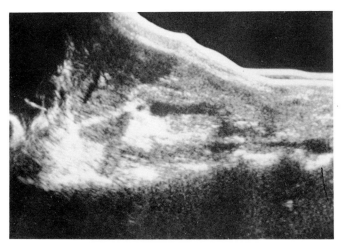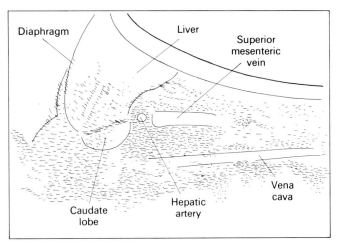

Fig. 11.39 Sagittal section. A short section of the superior mesenteric vein is outlined below the liver in the paramedian plane. It joins the splenic vein to form the portal vein.

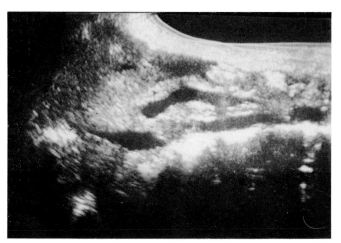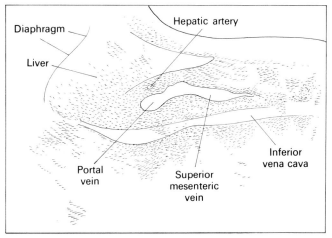

Fig. 11.40 Sagittal section. The superior mesenteric vein joins the splenic vein to form the portal vein anterior to the inferior vena cava.

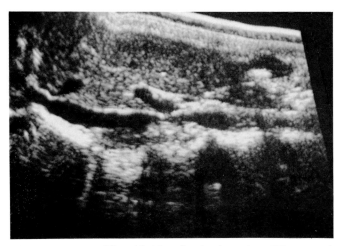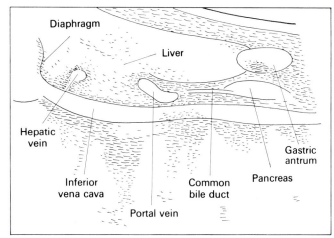

Fig. 11.41 Sagittal section. The portal vein lies anterior to the inferior vena cava above the pancreas and behind the liver. It is dividing into right and left branches.

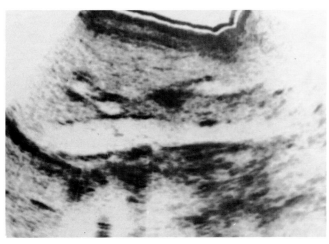 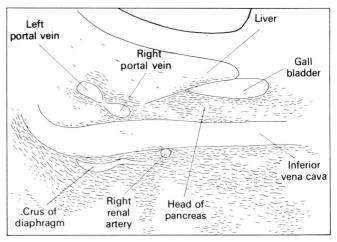

Fig. 11.42 Sagittal section. The portal vein has divided into its two main branches, the more anterior left portal vein and the transverse right portal vein.

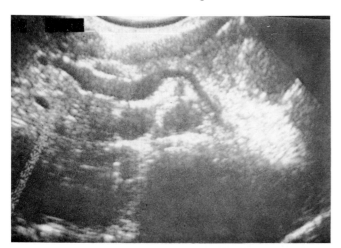 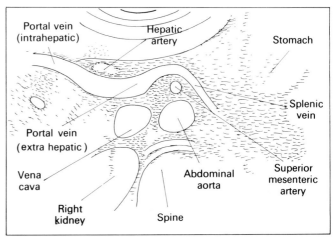

Fig. 11.43 Oblique transverse section. The splenic vein passes anterior to the superior mesenteric artery to form the portal vein anterior to the inferior vena cava. The portal vein passes upwards to the porta hepatis in the free edge of the lesser omentum. In the porta hepatis it lies behind the hepatic artery. The right portal vein is the direct continuation of the main portal vein in the liver.

VESSELS IN THE LOWER LIMB AND NECK

The larger vessels in the lower limb and the neck are visualised with conventional static B scanners, specially designed real-time systems and pulsed wave Doppler units. The Doppler units produce a visual image of the lumen of the vessel similar to the conventional arteriogram and the conventional B scanning systems outline the vessel wall.

The *femoral artery* is a continuation of the external iliac artery. It begins behind the inguinal ligament at the midinguinal point and passes down the anterior and medial aspect of the thigh. The upper part of the artery is in the femoral sheath contained in the femoral triangle, the lower part of the artery is in the subsartorial canal. It ends at the junction between the lower and middle third of the thigh where it becomes the popliteal artery. There are several branches supplying the inferior abdominal wall, the perineum and the skin, the fascia and the adductor, extensor and hamstring muscles of the thigh. The *profunda femoris* (deep femoral artery) arises about 3.5 cm below the inguinal ligament from the lateral or posterior aspect of the femoral artery, running initially posterior to the femoral artery to the medial side of the femur, to end in the lower third of the thigh by anastomosing with the superior muscular branch of the popliteal artery.

The *popliteal artery*, the continuation of the femoral artery, commences at the opening in adductor magnus at the lower end of the subsartorial canal. The artery passes through the popliteal fossa slightly lateral to the intercondylar fossa of the femur to divide into the anterior and posterior tibial arteries.

The *femoral vein* accompanies the femoral artery as the continuation of the popliteal vein and ends at the level of the inguinal ligament to become the external iliac vein. In its lowest part it is posterolateral to the femoral artery but passes behind and medial to the artery in the femoral triangle.

The *right common carotid artery* begins at the bifurcation of the innominate artery behind the sternoclavicular joint. The *left common carotid artery* arises from the highest part of the arch of the aorta to the left of the innominate artery. In the neck the *common carotid artery* passes obliquely upwards from behind the sternoclavicular joint to the upper border of the thyroid cartilage where it branches into the internal carotid and the external carotid artery. When it divides the vessel is slightly dilated forming the *carotid sinus* which also usually involves the proximal part of the internal carotid artery. The carotid artery is surrounded by the carotid sheath, a condensation of the deep cervical fascia, which also encloses the internal jugular vein and the vagus nerve.

The *external carotid artery* begins opposite the upper border of the thyroid cartilage and curves upwards and forwards behind the neck of the mandible. The *internal carotid artery* ascends to the base of the skull to enter the cranium through the carotid canal in the temporal bone. The *internal jugular vein* collects the blood from the brain and the superficial parts of the face and neck. It starts at the jugular foramen and runs down through the neck in the carotid sheath to unite with the subclavian vein to form the innominate vein. It lies lateral to the carotid artery.

In the neck the *vertebral vessels* lie in the foramina in the transverse processes of the upper six cervical vertebrae. The artery arises from the subclavian artery and the vein drains into the posterior aspect of the innominate vein.

VESSELS IN THE LOWER LIMB

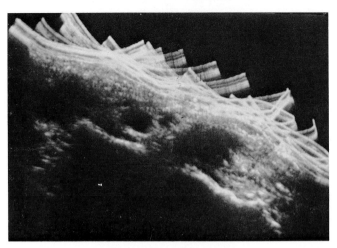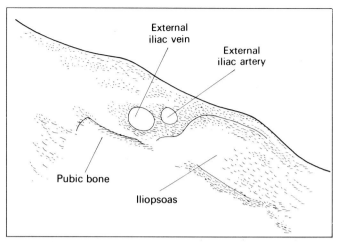

Fig. 11.44 Oblique transverse section above the inguinal ligament. The external iliac vein and artery lie anterior to the pubic bone and medial to the iliopsoas muscle.

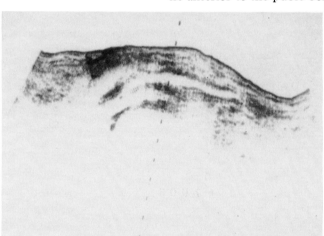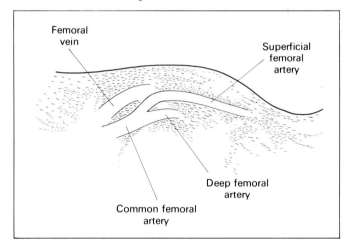

Fig. 11.45 Oblique sagittal section. The femoral vein and artery are in the femoral triangle just below the inguinal ligament. The femoral artery continues inferiorly as the superficial femoral artery after the origin of the deep femoral artery from its posterolateral aspect.

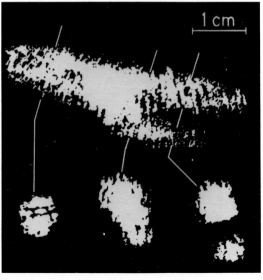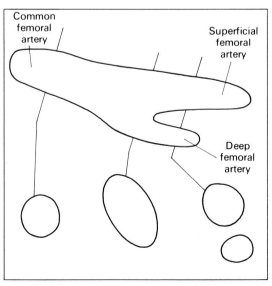

Fig. 11.46A Doppler scan. Sagittal (top) and transverse (lower) sections of the common femoral, superficial femoral and deep femoral arteries.

THE MAJOR VESSELS 195

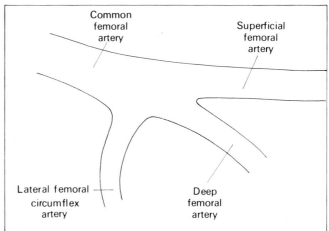

Fig. 11.46B Doppler scan. Sagittal section of the femoral and profunda arteries. The lateral femoral circumflex artery which usually arises from the deep femoral artery (profunda) arises from the common femoral artery in this subject.

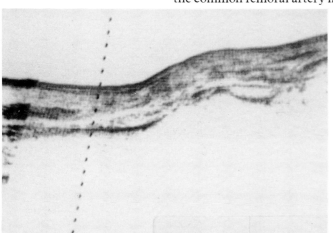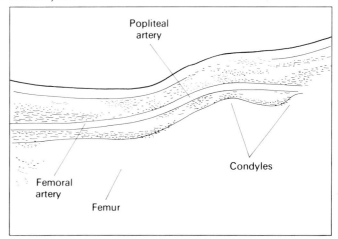

Fig. 11.47 Sagittal section. The femoral artery ends in the subsartorial canal at the junction of the lower and middle thirds of the thigh. It continues as the popliteal artery passing behind the condyles of the femur and tibia in the popliteal fossa.

VESSELS IN THE NECK

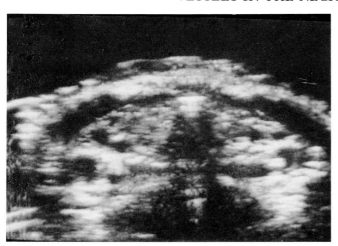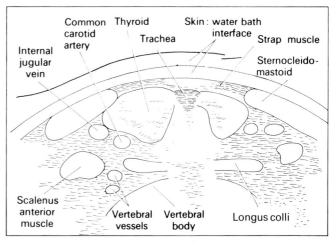

Fig. 11.48 Transverse section, water-bath technique, at the level of the thyroid gland. The common carotid arteries and jugular veins are lateral to and behind the lobes of the thyroid. At this level they are surrounded by the carotid sheath with the vagus nerve which is sometimes also outlined. The vertebral vessels are seen posteriorly on the left side between the prevertebral muscle (longus colli) and scalenus anterior.

196 ULTRASONIC SECTIONAL ANATOMY

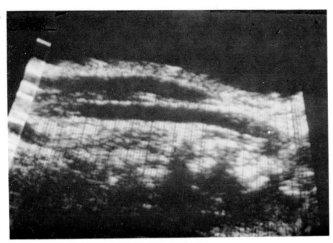 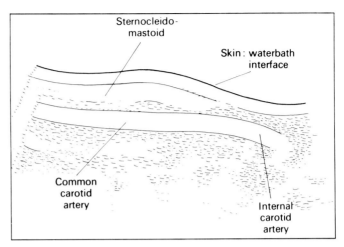

Fig. 11.49 Longitudinal section, water-bath technique. The common carotid and internal carotid artery are outlined. The external carotid is not demonstrated. (Scan reversed with the subject's head to the right of the section.)

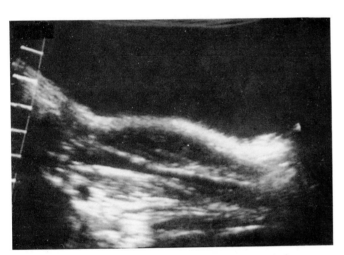 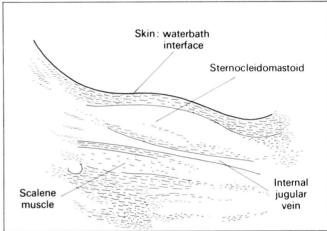

Fig. 11.50 Longitudinal section, water-bath technique. The internal jugular vein is lateral to the carotid artery. It lies behind the sternomastoid muscle. (Scan reversed with the subject's head to the right of the section.)

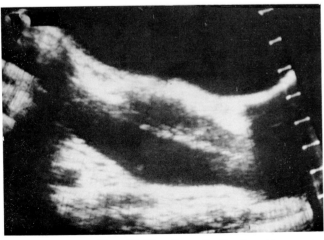 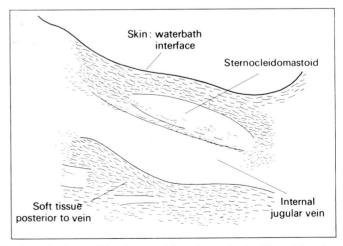

Fig. 11.51 Longitudinal section, water-bath technique. The internal jugular vein is dilated by the Valsalva manoeuvre. (Scan reversed with the subject's head to the right of the section.)

THE MAJOR VESSELS 197

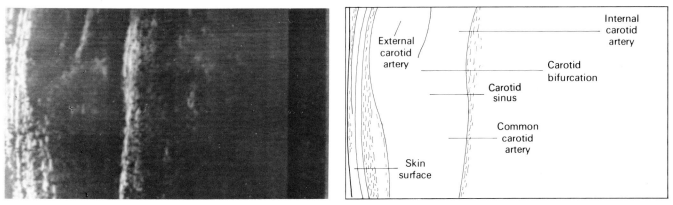

Fig. 11.52 Longitudinal section displayed as though the subject is in the erect position. Contact scan with small parts scanner. The bifurcation and carotid sinus are outlined.

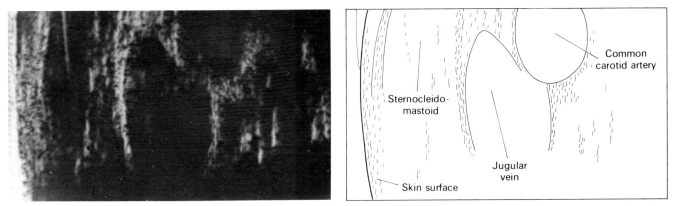

Fig. 11.53 Transverse section. Contact scan with small parts scanner. The jugular vein is lateral to the common carotid artery.

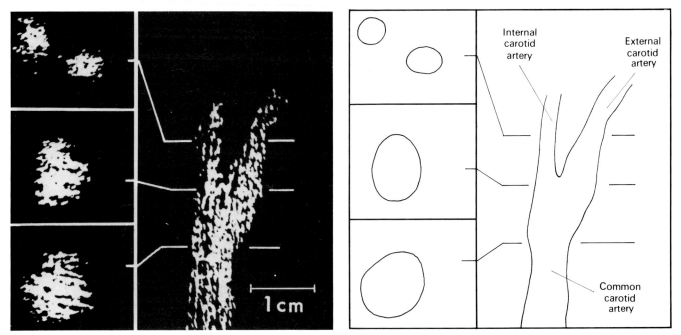

Fig. 11.54 Doppler scan of the common carotid artery and the internal and external branches. Transverse sections left and longitudinal section right displayed as though the subject is in the erect position.

12 OBSTETRIC ULTRASOUND

H. P. Robinson

The 'golden rule' of any ultrasound examination is to scan systematically in two places the whole volume of the 'area' of interest lest a misdiagnosis is made simply due to a failure of technique. This 'rule' is especially applicable in obstetrics where the glaring and most frequent error is to miss a second twin (or triplet). Given a systematic approach such errors should not be made.

The most useful preparatory action which may be taken prior to scanning a pregnant patient is to have her at least partially fill her bladder. This step is particularly important in the first trimester of pregnancy since the full bladder pulls the early gravid uterus up out of the pelvis, pushes away gas-filled intestines and acts as an excellent landmark for the easier identification and delineation of adjacent structures (Donald 1963). In later pregnancy where the imposition of bowel between the probe and the uterus has ceased to be a significant problem the partially-filled bladder continues to be useful as a landmark particularly when placenta praevia is suspected. In addition it may help to elevate the head when a BPD (biparietal diameter) measurement is difficult due to early engagement. Notwithstanding these positive uses of the full bladder technique an overfilled bladder may itself cause problems. By compressing the anterior and posterior uterine walls in the area of the lower segment the inferior edge of a normally sited placenta may appear to reach the 'internal os'. Similarly, in early pregnancy the whole uterus may be compressed causing elongation of the gestation sac with resulting difficulty in delineating its borders, and difficulty in visualising the fetus if it is trapped at the bottom of the sac.

The problem of supine hypotension in late pregnancy is a vexing one, and one which is not always controlled by tilting the patient to the left and/or by elevating the lower end of the examining table. Because of its serious import, failure to alleviate this problem should lead to abandonment of the examination.

While obstetric scanning is frequently an easy procedure it can be difficult, time-consuming and demanding of patience and clinical insight.

BASIC EMBRYOLOGY

At approximately mid-cycle the ovarian follicle ruptures and the ovum, with adherent cells from the cumulus ovaricus, escapes into the peritoneal cavity. From there the ovum passes into the fimbriated end of the fallopian tube, which at the time of evolution is closely applied to the ovary.

Fertilisation takes place in the fallopian tube. The fertilised ovum then travels down the tube for implantation into the uterus. As it passes down the tube the single-celled ovum commences to divide forming a small spherical mass called the *morula*. Between the third and seventh day after fertilisation a cavity forms in the morula, forming the *blastocyst*. The early blastocyst consists of a *formative mass* and a ring of *trophoblast*. The formative mass differentiates to form the *germ disc*, the *primitive entoderm* and the *primary mesoderm* which grows around to line the inner surface of the trophoblast. At the stage of implantation into the uterine wall the blastocyst consists of an inner embryonic cell mass lying at one pole of the cavity and a spherical envelope forming trophoblast and mesoderm. Implantation is usually on the posterior wall of the uterus, nearer to the fundus than to the cervix, and may be in the median plane or to one side.

The trophoderm erodes the epithelium of the endometrium and becomes embedded about the 14th day after fertilisation which is approximately at four weeks gestation—the period of gestation being estimated from the first day of the last menstrual period. The cells of the trophoblast penetrate the maternal epithelial and stromal tissue which has undergone changes in preparation for implantation. These changes are stimulated by the progesterone of the corpus luteum which persists if fertilisation of the ovum takes place. The trophoblast undergoes rapid proliferation forming the *trophoblastic shell*. Initially the trophoblastic cells digest the endometrium, then *chorionic villi* develop and the maternal vessels open up to form a lake of maternal blood. By the beginning of the third week after fertilisation (five weeks gestation) the blastocyst has an outer layer of chorionic villi lined by primary mesoderm; the cavity of the blastocyst is the *extraembryonic coelom* and the embryonic area which lies at one pole of the blastocyst has developed a *yolk sac*, an *embryonic disc*, *amniotic cavity* and *connecting stalk*, between the tail of the embryonic disc and the trophoblast. This represents the stage of development between the fourth and fifth weeks of gestation and at this time the developing blastocyst is just definable on an ultrasonic section. The embryonic area or *fetal pole* is however not defined.

The decidua

Before a fertilised ovum reaches the uterus the mucous membrane of the uterus must be ready for its reception. Following the end of the menstrual flow a reparative process of the mucous membrane takes place lasting actively for about seven days. If fertilisation of an ovum occurs the *corpus luteum* continues to function actively and the thickened vascular mucous membrane known as the *decidua* becomes ready for reception of the fertilised ovum. The interglandular tissue increases in quantity and is crowded with decidual cells. After the ovum is embedded distinctive names are applied to the different parts of the decidua. That which covers the ovum is known as the *decidua capsularis*. The portion between the ovum and the uterine muscular wall is termed the *decidua basalis*, and it is from here that the placenta develops. The decidua lining the remainder of the uterus is called the *decidua parietalis*. By the second month of pregnancy the mucous membrane consists of three strata: the stratum compactum, next to the free surface; the stratum spongiosum where the uterine glands are dilated and tortuous, and the limiting or boundary layer next to the uterine muscular fibres. With the growth of the embryo and expansion of the amniotic cavity the decidua capsularis becomes thinned and gradually comes in contact with the decidua parietalis. By the fifth month of gestation the decidua capsularis has practically disappeared and in the succeeding months the decidua parientalis gradually thins leaving a thin layer of cells—the chorion laeve.

Formation of the embryo

Until the end of the third week after fertilisation (the fifth week of gestation) the embryonic area is disc-like, but at the end of this week the embryo begins to form its definitive shape. The head fold and tail fold develop and lateral folds gradually constrict the yolk sac to give it its characteristic shape. Before the formation of the tail fold the embryonic area is anchored to the trophoblast by a *connecting stalk*. When the tail fold develops this stalk is carried round onto the ventral aspect of the embryo to assume the permanent position of the umbilical cord. The *yolk sac* becomes included in both head and tail folds to form the foregut and hind-gut. The intervening dorsal portion of the yolk sac constitutes the mid-gut. At first the mid-gut communicates freely with the rest of the yolk sac but the connection gradually becomes drawn out to form the *vitello-intestinal duct*. The remainder of the yolk sac remains extraembryonic and is often termed the *umbilical vesicle*.

The fetal membranes and placenta

The allantois forms about the fifth week of gestation as an outgrowth of the yolk sac. Initially solid, it soon becomes canalised and becomes incorporated in the hind gut. This allantois is lined with entoderm from which the umbilical vessels develop.

The amnion is a membranous sac surrounding the embryo which develops as a cavity between the formative mass and the trophoblast. Fluid termed *liquor amnii* occupies the cavity increasing steadily, and the sac gradually expands to encroach on the extra-embryonic coelom which is finally obliterated except for a small portion included in the umbilical cord. The amniotic fluid increases in volume up to the sixth or seventh month of pregnancy; at term it amounts to about one litre.

The umbilical cord

The *connecting stalk* is a mass of primary mesenchyme which connects the tail end of the embryonic area with the chorion. With embryonic development it comes to lie on the ventral surface of the embryo and its mesenchyme approaches the yolk sac and the vitello-intestinal duct. The lateral folds of the amnion converge on the region of the connecting stalk and vitello-intestinal duct to form the *umbilical cord*. The cord consists of an outer covering of amnion, containing the vitello-intestinal duct and a mass of primary mesenchyme. It incorporates the connecting stalk and its umbilical vessels and the allantoic canal. The part of the extra-embryonic coelom incorporated in the cord acts as a sac for the normal umbilical hernia which characterises the embryo between the 8th and 10th week of gestation. It normally becomes obliterated after this stage. The cord becomes spirally twisted and elongates so that by term it is 50 cm long.

The development of the placenta

As already stated the fertilised ovum adheres to the uterine mucosa, destroys the epithelium and excavates a cavity in the mucous membrane. The structure concerned with excavation is the plasmodial trophoblast. The *chorion* consists of an outer layer of trophoblast and an inner layer of primary mesenchyme. It undergoes rapid proliferation forming *primary*, *secondary* and *tertiary chorionic villi*. The trophoblast invades and digests the uterine tissue and attacks the walls of the maternal vessels. Lacunar spaces develop in the trophoblast and enlarge, their walls becoming sponge-work and forming the primary chorionic villi, from which secondary and tertiary villi develop. From the fifth week of gestation the entire chorion is covered with villi projecting into the decidua capsularis and basalis. With the growth of the embryo the decidua capsularis thins out and the corresponding chorionic villi atrophy. The trophoblast in contact with the decidua basalis increases in size to form the *chorion frondosum*—it constitutes the placental area.

The placenta connects the fetus to the uterine wall. The fetal part of chorion frondosum branches repeatedly and increases enormously in size. Some branches are anchored by columns of trophoblasts to the walls of the intervillous spaces but the majority hang free in space. All are bathed in maternal blood. Blood is carried by the umbilical arteries from embryo to placenta and returns by the umbilical vein.

The maternal portion of the placenta is formed by the decidua basalis containing *intervillous spaces*. The formation of the intervillous space involves the disappearance of the stratum compactum of the decidua, but the deeper layer of this remains to form the basal plate which constitutes the outer wall of the intervillous space. The interval between the basal plate and the uterine muscle is occupied by a compressed and atrophied stratum spongiosum and limiting layer. Through these and the basal plate the uterine arteries and veins pass to and from the intervillous space. Portions of the stratum compactum persist as pillars which project into the intervillous spaces which later form the *placental septa* which incompletely divide the placenta into *lobes* or *cotyledons*. The fetal and maternal blood do not mix, being separated from each other by the delicate walls of the villi. The placenta acts as a mechanical connection between the mother and fetus as well as acting as an organ of respiration, nutrition and excretion for the fetus.

THE FORM OF THE EMBRYO AT DIFFERENT STAGES OF GROWTH RELATED TO THE ULTRASONIC IMAGE

Gestational age

1. 3 weeks — Initially adheres to endometrium, followed by implantation in the decidua. At early blastocyst stage.
2. 4 weeks — Blastocyst containing amnio-embryonic and yolk sac vesicles with an embryonic area lying between the two vesicles. *Blastocyst may be identified in uterus.*
3. 5 weeks — Differentiation of embryo commences. *Blastocyst identified as gestation sac.*

4. 6 weeks   Head and tail folds completed. Neural fold begins to close. Primitive limb buds present. Heart begins to beat. *Fetal echoes identified in gestation sac.*
5. 7th week   Curvature of embryo increases. Limb buds lengthen, hands and feet recognised. *Fetal heart movements recorded. C.R.L. (crown rump length) measurements used to date fetus.*
6. 8th week   Further curvature of embryo. U loop of gut extruded from abdomen. *Thick ring of trophoblast. Fluid in gestation sac now represents amniotic fluid.*
7. 9th and 10th weeks   Flexure of head reduced and neck lengthens. Fingers and toes recognisable. *Differentiation of head and trunk possible. Placental site defined. Decidua capsularis thinning.*
8. 11th week Embryo now in fetal period. *Placenta differentiating. Fetal detail developing.*
9. 12th week Head extended. Limbs developed. Umbilical hernia reduced. *Cerebral midline seen. B.P.D. measurements possible. Good fetal detail. Fetal spine defined.*
10. 14 weeks *Placenta now fully developed with chorionic plate. Clear fetal detail. Cerebral ventricular system seen. Fetal movement present.*
11. 16 weeks Fetal anatomy defined. B.P.D. measurements used to date fetus.

Editorial note: Though not correct embryologically the term fetus is used by sonographers when referring to the developing embryo prior to the fetal stage.

THE FIRST TRIMESTER

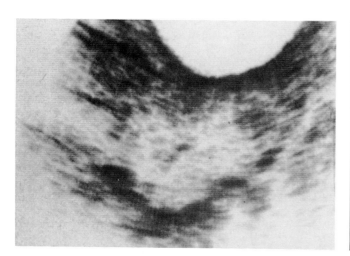 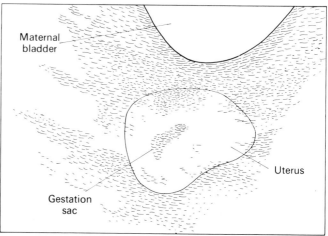

Fig. 12.1 Transverse section through the uterine fundus at four weeks gestation. The fertilised ovum is in the blastocyst stage and is embedded in the uterine decidua. This is the first ultrasonic evidence of gestation and is seen as a strong endometrial echo; the central fluid in the blastocyst cannot be demonstrated at this stage of gestation.

OBSTETRIC ULTRASOUND 203

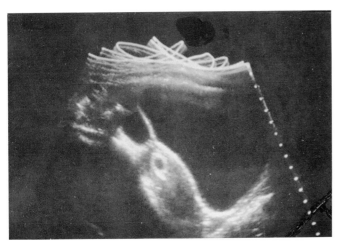 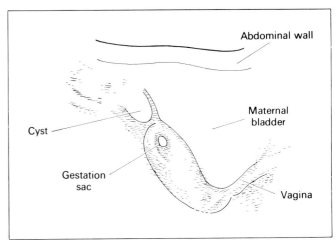

Fig. 12.2 Sagittal section at 4+ weeks gestation. The blastocyst has continued to enlarge and is seen as a central sonolucency surrounded by a strongly echogenic ring. This is known as the gestation sac. The ring consists of the trophoblastic layer of chorion, and it is the chorionic villi that form the ring. The central fluid at this stage represents the extraembryonic coelom. The disc-like embryonic area cannot be seen. There is a small retention cyst above the uterine fundus. (See Fig. 12.16). (Scale 2:5)

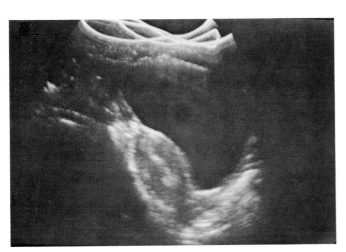 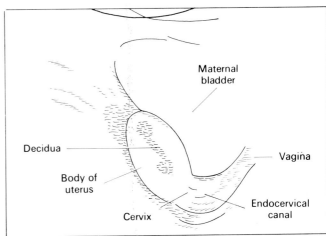

Fig. 12.3 Sagittal section in same subject as 12.2. The thickened vascular mucosa or decidua parietalis lines the cavity of the uterus. (Scale 2:5)

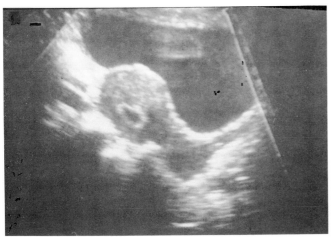 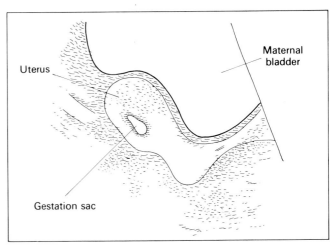

Fig. 12.4 Sagittal section at six weeks gestation. The sac has increased in size but fetal echoes are not yet demonstrated. (Scale 2:5)

204 ULTRASONIC SECTIONAL ANATOMY

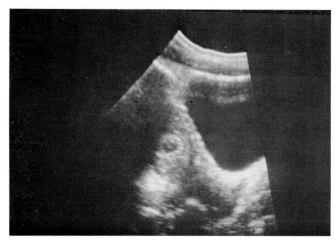 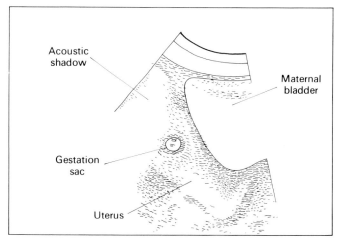

Fig. 12.5A Sagittal section. The gestation period is between 6 weeks and 7 weeks. The chorionic villi forming the outer ring (trophoblastic shell) have increased in thickness. Central echoes in the sac represent the developing fetus. (Scale 2:5)

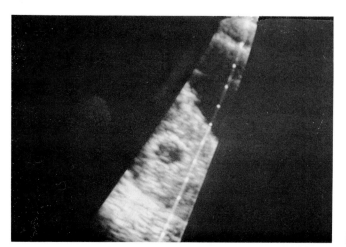 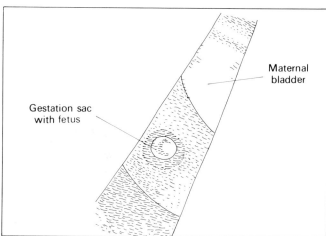

Fig. 12.5B Limited sector scan of gestation in 12.5A (Scale 3:5)

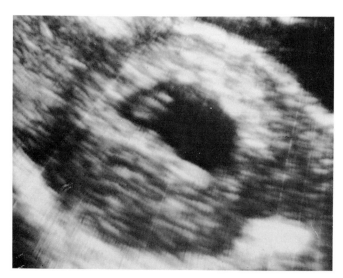 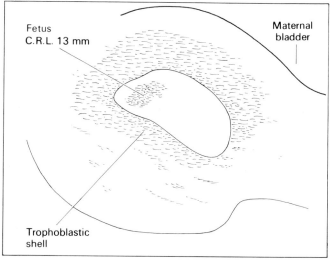

Fig. 12.6 Limited field section at 6 to 7 weeks gestation. The fetus is defined and the C.R.L. measures 13 mm. (Crown rump length measurements are used to date a gestation from the 7th to the 15th week). (Scale 5:5)

OBSTETRIC ULTRASOUND 205

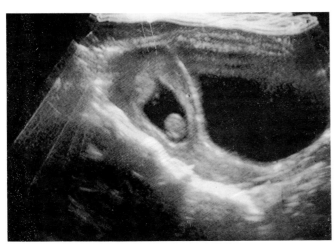 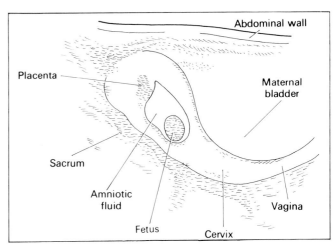

Fig. 12.7 Sagittal section. Gestation period at 9+ weeks. The fetus has grown and the sac has increased in size with the development of the amniotic cavity. The placenta is developing in the region of the decidua basalis; the decidua capsularis is thinning. (Scale 2:5)

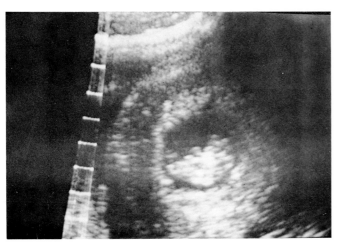 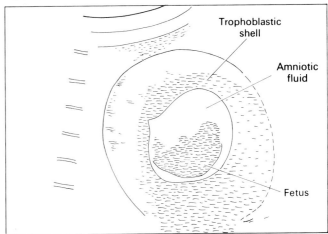

Fig. 12.8 Limited field section at 9+ weeks gestation. The C.R.L. is between 26 mm and 28 mm at 9 weeks. (Scale 4:5)

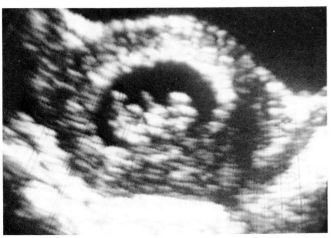 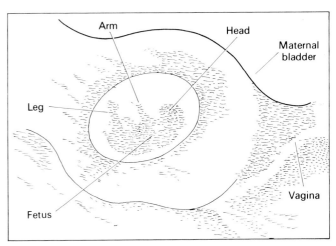

Fig. 12.9 Limited field section. Fetal detail at 10 weeks gestation. The C.R.L. is 29 mm and the head and trunk can now be differentiated. (Scale 5:5)

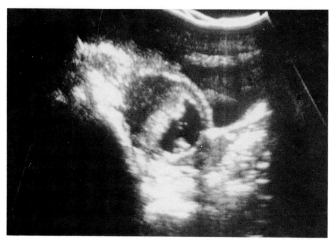 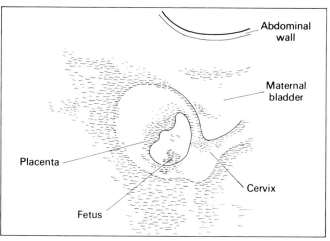

Fig. 12.10 Sagittal section at 11 weeks gestation. The placenta is developing at the site of the decidua basalis. The amniotic cavity has increased in size and the decidua capsularis and adjacent trophoblast is thinning as a result of normal atrophy. The fetus is not fully displayed on this section. (Scale 2:5)

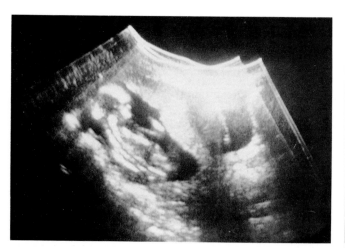 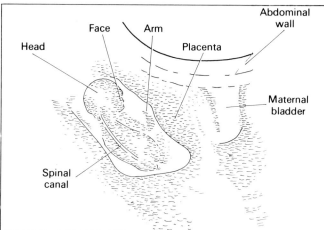

Fig. 12.11 Limited field scan. Fetus at 13 weeks gestation. Clear fetal detail is seen. At this stage it is possible to obtain B.P.D. and C.R.L. measurements. B.P.D. measurements start at 12 weeks gestation but are more accurate from 16 weeks.

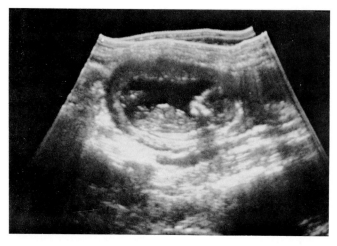 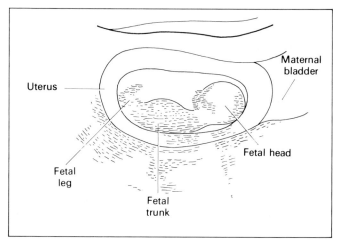

Fig. 12.12 Fetus at 14 weeks gestation. (Scale 2:5)

OBSTETRIC ULTRASOUND 207

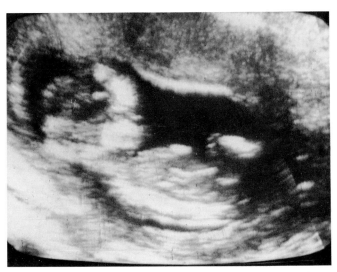 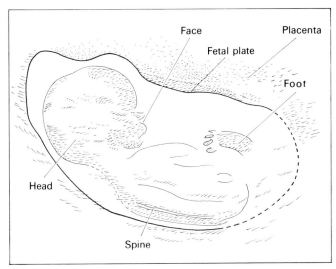

Fig. 12.13 Limited field scan. Fetus at 14 weeks gestation. The C.R.L. is 88 mm. The fetus is clearly seen in the amniotic cavity. The placenta lies anteriorly with the chorionic or fetal plate evident. (Scale 5:5)

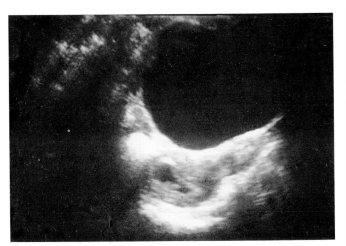 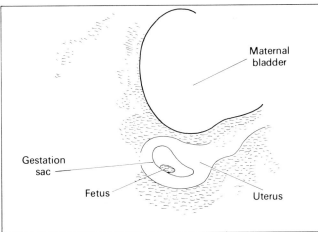

Fig. 12.14 Sagittal section. Retroverted uterus with 8 weeks gestation. Detail is not so clearly seen as with an anteverted uterus and an early fundal implantation can be overlooked. (Scale 5:5)

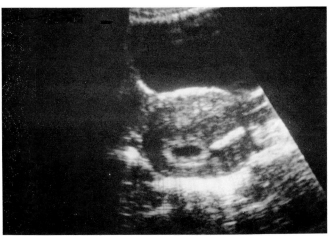 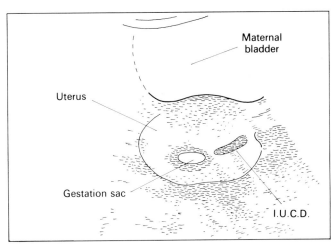

Fig. 12.15 Transverse section. Early pregnancy with an associated IUCD adjacent to the gestation sac. (Scale 3:5)

208 ULTRASONIC SECTIONAL ANATOMY

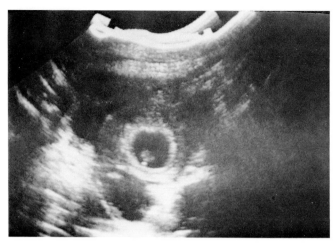 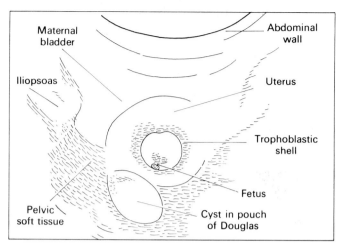

Fig. 12.16A Transverse section with caudal angulation. Eight weeks intra-uterine gestation with a small retention cyst in the pouch of Douglas. These cysts are usually corpus luteum cysts and most regress spontaneously. (Scale 2:5)

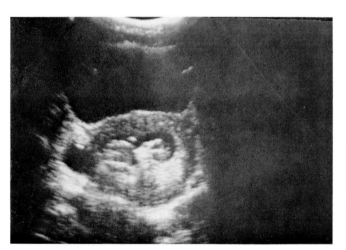 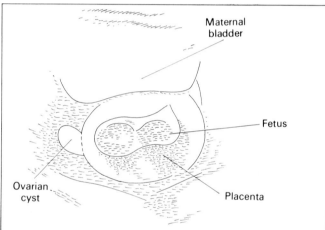

Fig. 12.16B Transverse section. Fourteen weeks intrauterine gestation with small retention cyst of the right ovary. (Scale 2:5)

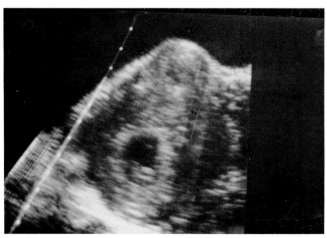 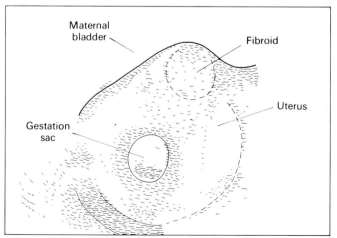

Fig. 12.17A Transverse section. Seven weeks gestation with a small anterior fibroid. (Scale 3:5)

OBSTETRIC ULTRASOUND 209

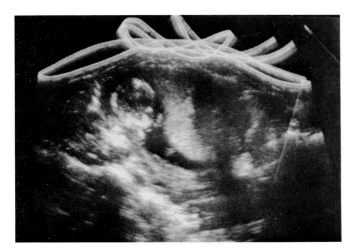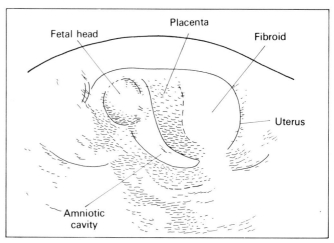

Fig. 12.17B Sagittal section. Sixteen weeks gestation with an anterior fibroid below the placenta. (Scale 2:5) There was no change on serial observation.

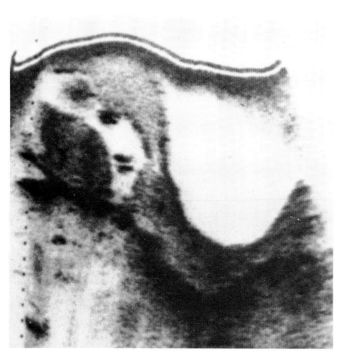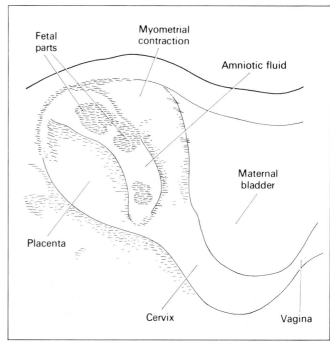

Fig. 12.18 Sagittal section. Intrauterine pregnancy. An area of myometrial contraction is seen anteriorly. These usually occur in the second trimester and simulate uterine fibroids or may be mistaken ultrasonically for the placenta. Observation over a period of time will demonstrate that these areas of myometrial thickening are transient.

TWIN GESTATION

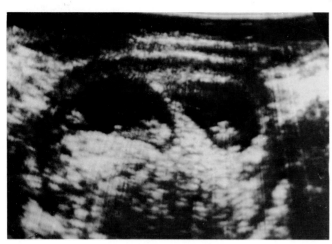 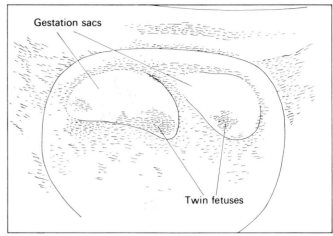

Fig. 12.19 Limited transverse section at 8 weeks gestation. Twin sacs are present separated by a curvilinear septum. A fetus is present in each sac. Multiple pregnancy can be diagnosed as early as 7 weeks gestation. (Scale 3:5)

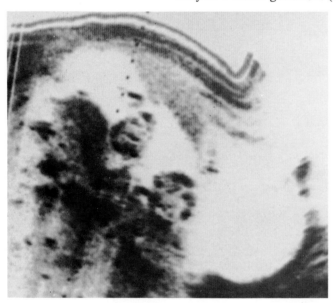 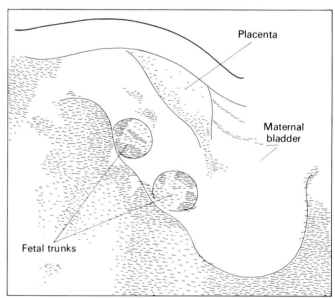

Fig. 12.20 Sagittal section. Twin pregnancy at 13 weeks. Both trunks are seen on the one section. Only one placenta could be identified.

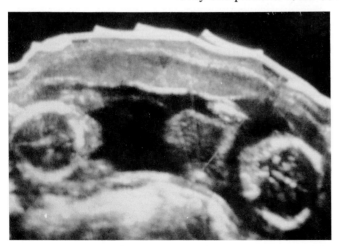 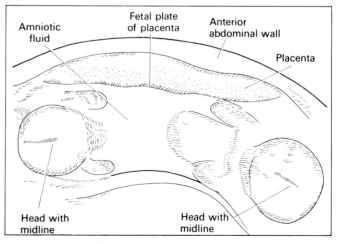

Fig. 12.21 Sagittal section. Twin pregnancy at 22 weeks. Two heads are shown on the one section. There was only one placenta. (Scale 2:5)

OBSTETRIC ULTRASOUND 211

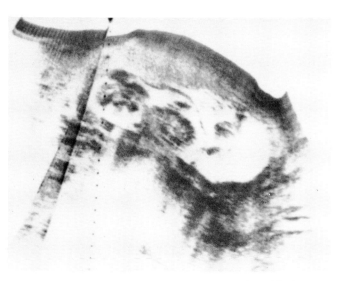 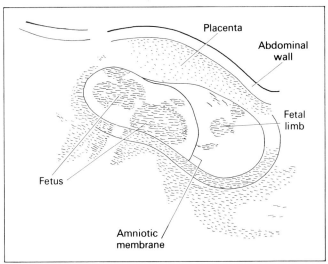

Fig. 12.22 Sagittal section. Twin pregnancy with an amniotic membrane separating the cavities. Only one placenta was identified.

 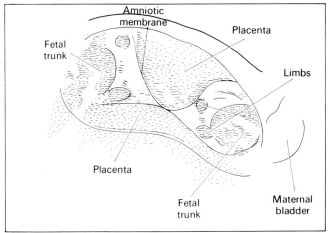

Fig. 12.23 Sagittal section. Twin gestation with twin placentas. Two fetal trunks are present. One placenta is posterior, the second is attached to the amniotic membrane.

FETAL ANATOMY

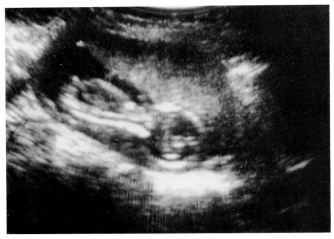 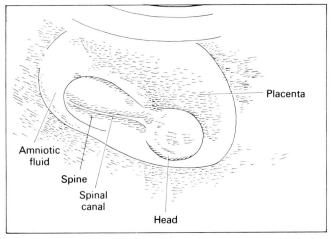

Fig. 12.24 Longitudinal section of fetal spine at 12 weeks gestation. The spine is seen as a tubular structure extending from the base of the skull to the sacral region. The fetal spinal canal is relatively wide and is defined between the echogenic vertebrae.

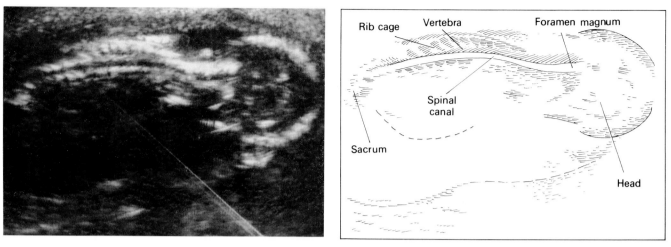

Fig. 12.25 Longitudinal section of fetal spine at 16 weeks gestation. In this section the cervical and dorsal spine is well demonstrated. The ribs are also seen in the dorsal region.

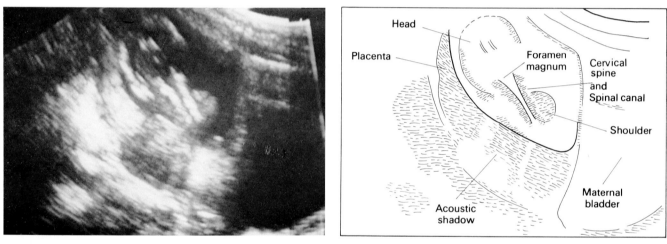

Fig. 12.26 Longitudinal section of the cervical spine, with the foramen magnum at the base of the skull.

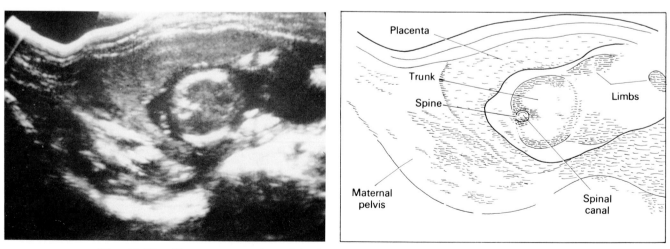

Fig. 12.27 Transverse section of fetal trunk at 14 weeks gestation. The spine is seen as a circular echogenic structure. The spinal canal is central and is surrounded by the echogenic vertebra.

OBSTETRIC ULTRASOUND 213

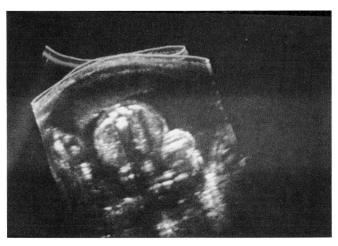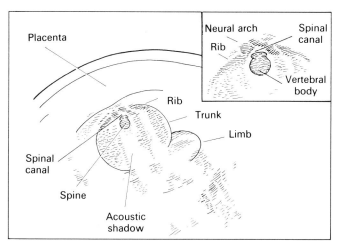

Fig. 12.28 Transverse section of fetal chest. Vertebral detail is clearly seen with the ribs demonstrated laterally. Both the spine and ribs are strongly echogenic and there is distal shadowing.

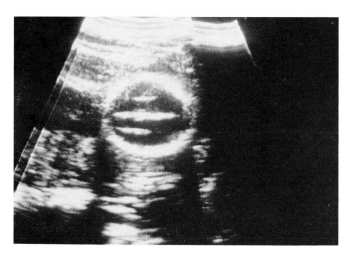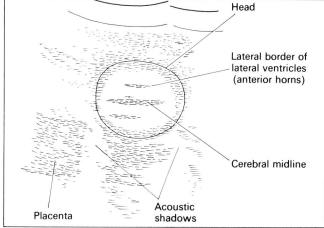

Fig. 12.29 Transverse section of fetal head at 16 weeks gestation with the cerebral midline and the lateral ventricles defined. The fetal head is seen as early as 10 weeks and the midline from 12 weeks onwards. The lateral walls of the lateral ventricles are recorded as lines parallel to the midline from early in the second trimester. The LVR is high at this stage of gestation (p. 8).

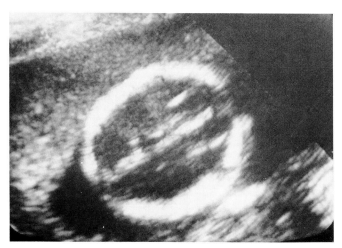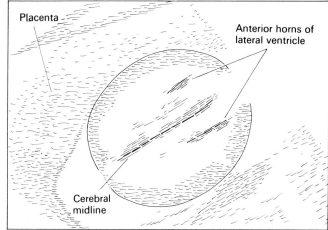

Fig. 12.30 Fetal head at 18 weeks gestation with the lateral walls of the anterior horns of the lateral ventricles seen on either side of the midline.

214 ULTRASONIC SECTIONAL ANATOMY

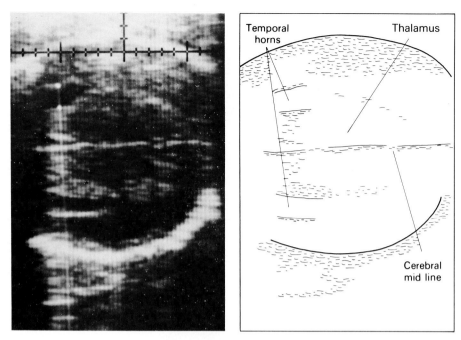

Fig. 12.31 Transverse section of fetal head recorded with a linear array. The temporal horns of the lateral ventricles are clearly seen. The cerebral ventricular system is easier to view with real-time systems than with conventional static B scanners. The thalamus, an important landmark for obtaining biparietal diameters, is well demonstrated.

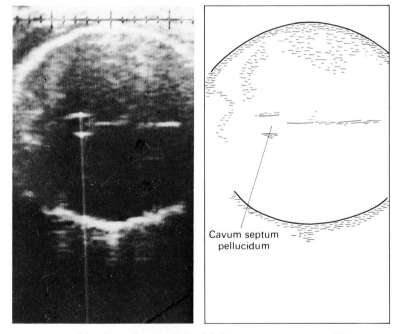

Fig. 12.32 Transverse section of fetal head recorded with linear array system. The two lamini of the septum pellucidum are separated by a cavity.

OBSTETRIC ULTRASOUND 215

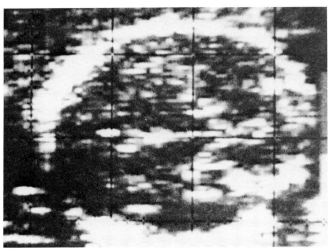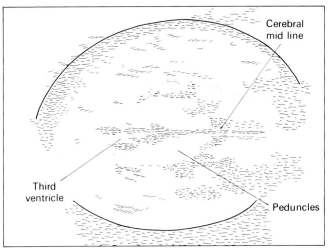

Fig. 12.33 Transverse section through the third ventricle and the cerebral peduncles. The cerebral peduncles emerge from the upper surface of the pons passing upwards and forwards into the cerebral hemisphere. Section recorded with linear array.

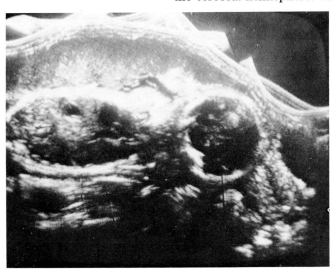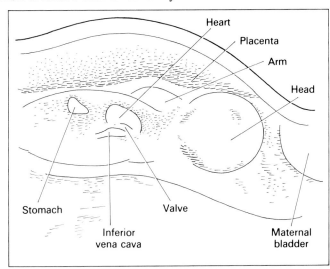

Fig. 12.34 Longitudinal section through the fetal head and trunk with the placenta lying anteriorly. The heart is indistinctly outlined in the thorax with the inferior vena cava posterior and inferior. The cardiac chamber and valves can be identified but are best appreciated with real-time systems. The fluid-filled stomach is seen below the diaphragm. Vertex presentation.

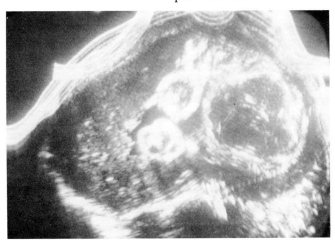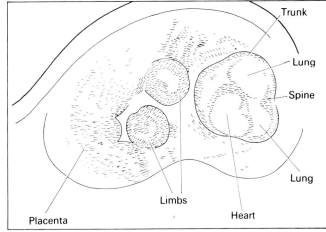

Fig. 12.35 Transverse section. The fetal thorax and limbs are outlined, the placenta lies laterally. The lungs are seen surrounding the heart; they are homogeneous in texture. The heart is outlined posteriorly, but intracardiac detail is not demonstrated. Two limbs are sectioned; the highly echogenic structures within the limbs represent the developing skeleton.

216 ULTRASONIC SECTIONAL ANATOMY

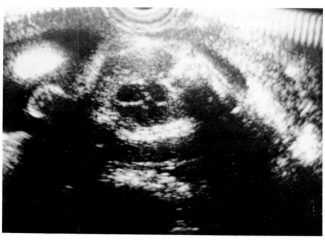 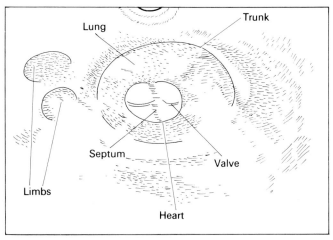

Fig. 12.36 Wide angle sector scan taken with a mechanic real-time system. The thorax is sectioned. Cardiac detail is demonstrated with the septum and valves identified. The texture of the fetal lungs is well defined. Fetal heart : fetal chest ratio is 0.52 (Garrett & Robinson 1970).

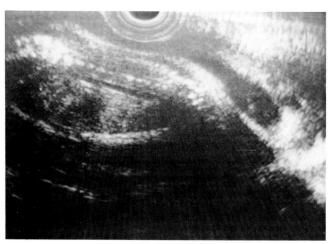 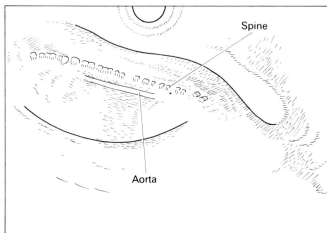

Fig. 12.37 Wide angle sector scan. The abdominal and thoracic aorta is outlined ventral to the fetal spine.

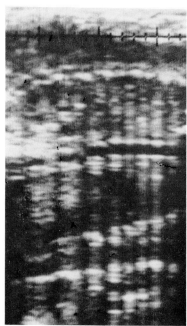 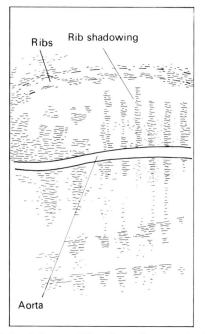

Fig. 12.38 Longitudinal section through fetal thorax. The thoracic aorta is clearly defined. There is marked acoustic shadowing from the ribs.

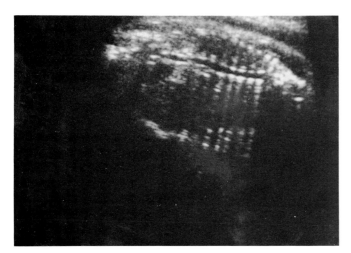 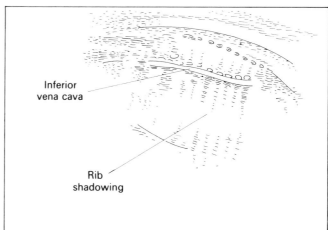

Fig. 12.39 Longitudinal section through the fetal abdomen and thorax. The inferior vena cava is outlined. Its appearance is similar to the aorta but the vena cava enters the posterior inferior aspect of the heart, whereas the aorta can be followed up into the thorax behind and above the cardiac outline to the level of the aortic arch.

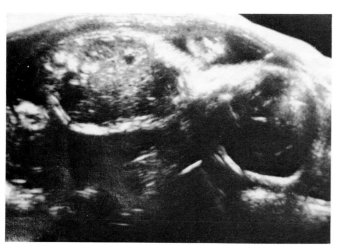 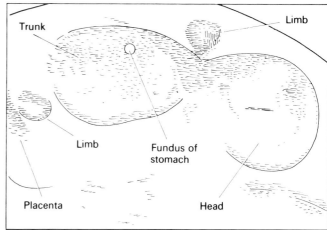

Fig. 12.40 Longitudinal section through fetal trunk and head. The fluid-distended fetal stomach is seen in the upper part of the abdomen. The stomach is commonly seen on both transverse and longitudinal sections lying above the level of the ductus venosus—its maximum transverse diameter is 25 mm (Garrett 1980).

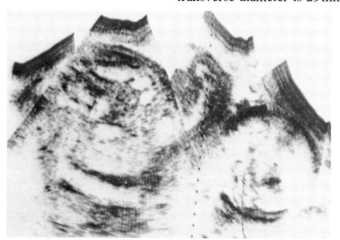 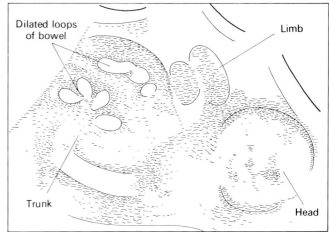

Fig. 12.41 Longitudinal section through fetal trunk and head. Multiple fluid-dilated loops of bowel are present in the abdomen. After delivery this fetus was shown to have a normal gastrointestinal tract.

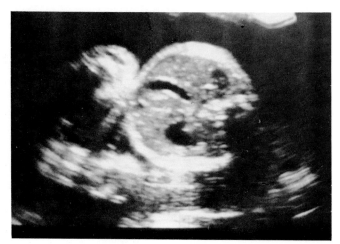 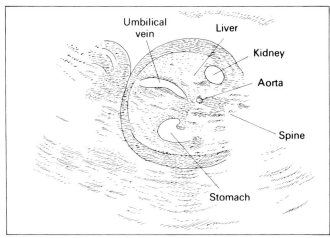

Fig. 12.42 Transverse section of the fetal abdomen at the level of the umbilical vein. Abdominal circumference measurements are taken at this level. The umbilical vein, which measures up to 1 cm in diameter, carries the blood from the placenta to the fetus and at the porta hepatis it joins the left portal vein. The ductus venosus, the continuation of the umbilical vein, originates from the left portal vein opposite this point and ascends to join the left hepatic vein immediately before it enters the inferior vena cava. When the placental circulation is cut off after birth, the umbilical vein becomes thrombosed and develops into a fibrous cord—the ligamentum teres. The ductus venosus forms the ligamentum venosum. In this section the homogeneous grey texture of the liver is well demonstrated. Its inferior surface is difficult to differentiate from collapsed bowel though the fluid-distended stomach is readily identified. The right kidney is clearly seen behind the liver adjacent to the spine and the abdominal aorta.

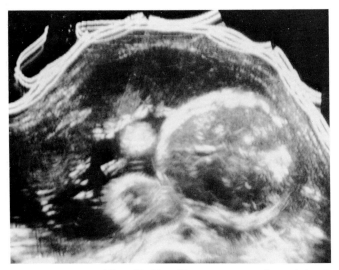 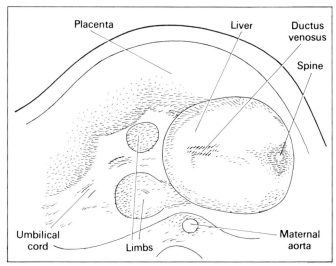

Fig. 12.43 Transverse section of the fetal abdomen at the level of the umbilical vein. The umbilical cord is seen in the amniotic fluid passing laterally to the placenta.

OBSTETRIC ULTRASOUND 219

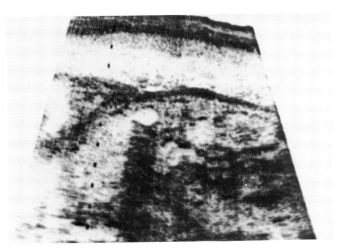 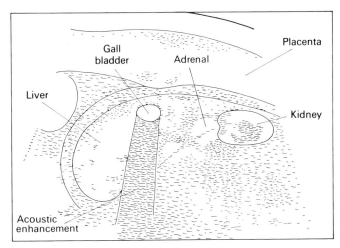

Fig. 12.44 Limited transverse section of fetal abdomen. The gall bladder is distended and is seen between the liver and right kidney.

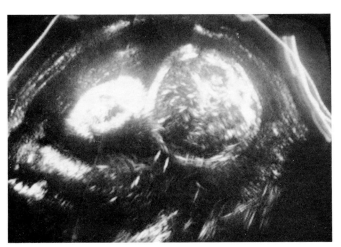 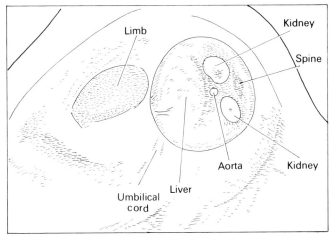

Fig. 12.45 Transverse section through the fetal abdomen. Both kidneys are seen posteriorly adjacent to the spine with the abdominal aorta anteriorly. The liver lies ventrally on the right of the abdomen extending over to the left side. The remainder of the abdominal cavity is filled with collapsed gut.

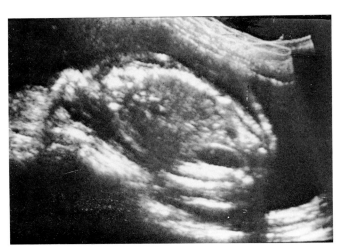 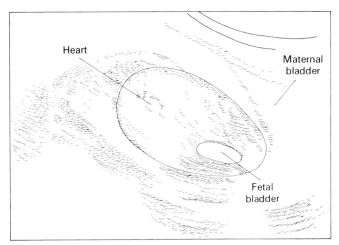

Fig. 12.46 Longitudinal section of fetal trunk. The urinary bladder is seen in the lower abdomen; its volume does not usually exceed 60 ml. The fetal bladder is seen to empty periodically: the hourly fetal urine production rate (HFUPR) can be estimated from measurement of fetal bladder diameters (Campbell et al 1973).

220 NORMAL ULTRASONIC SECTIONAL ANATOMY

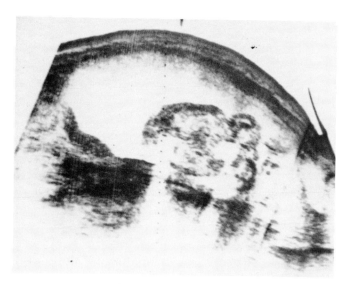 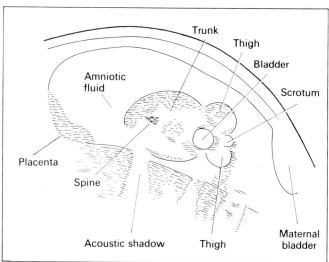

Fig. 12.47 Transverse section through the fetal bladder in a male fetus. In the fetus the bladder lies in the lower abdomen. This section also passes through the upper part of both thighs and the scrotum can be identified.

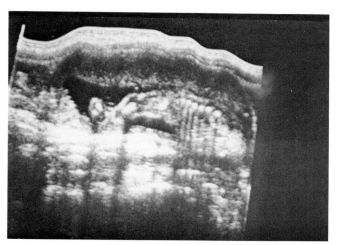 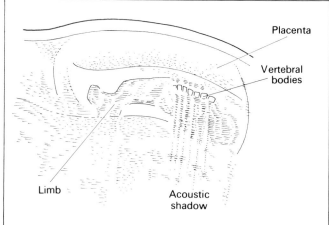

Fig. 12.48 Longitudinal section of a fetal limb. The developing skeleton is seen as highly reflective segments in the limb. The fetal spine is also seen, with the vertebral bodies defined.

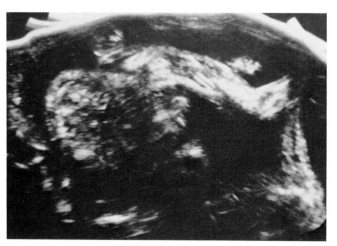 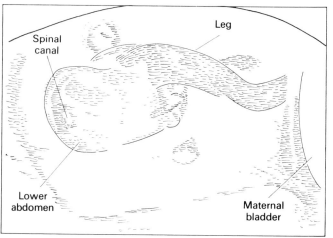

Fig. 12.49 Sagittal section. Limb presentation at 34 weeks gestation.

OBSTETRIC ULTRASOUND 221

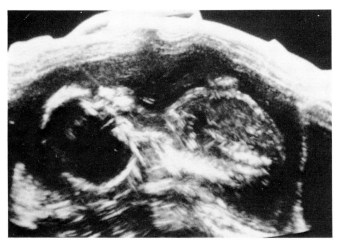 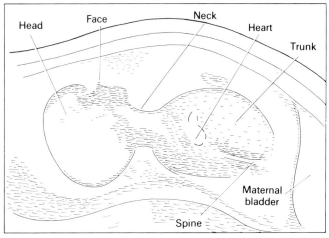

Fig. 12.50 Sagittal section. Breech presentation at 34 weeks gestation. (See Fig. 12.34 for vertex presentation).

THE PLACENTA

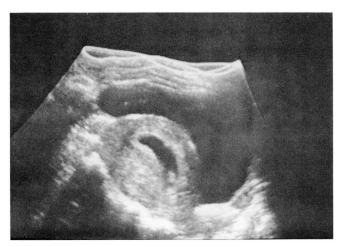 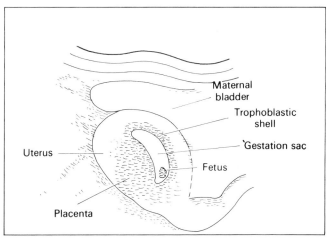

Fig. 12.51 Sagittal section. Nine weeks gestation with the site of the placenta defined. The placenta develops from the trophoblast at the site of the decidua basalis, which is in the portion of the decidua lying between the ovum and the uterine muscular layer. (Scale 2:5)

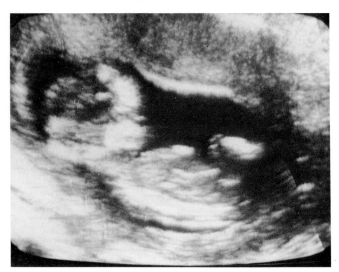 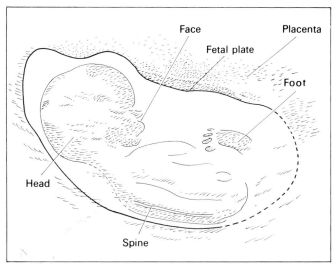

Fig. 12.52 Longitudinal section through mature placenta at 14 weeks gestation. The placenta is developed as a mature organ by the 12th to 16th week of pregnancy. The mature placenta is homogeneous with a ground glass appearance. The fetal surface or chorionic plate is smooth and is a defined layer which consists of amnion and chorion with umbilical vessels. This appearance is characteristic of a Grade 0 placenta and all normal first and second trimester placentas assume this appearance until approximately 28–30 weeks gestation. (Scale 5:5)

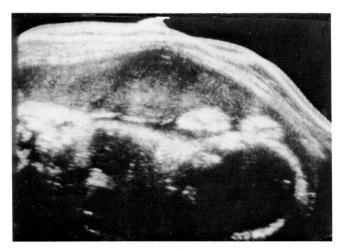 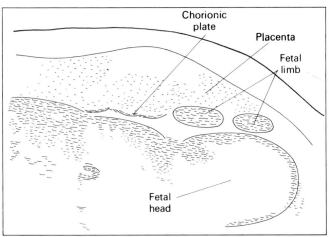

Fig. 12.53 Longitudinal section through a Grade 1 placenta. The chorionic plate assumes subtle indentations and small echogenic densities appear which are randomly dispersed through the placental substance except in the basal layer. These echogenic densities vary from 1 mm to 4 mm in size. (Scale 2:5).

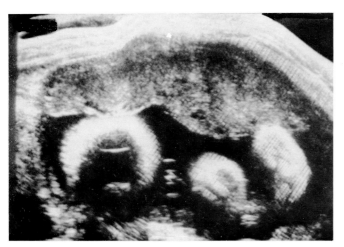 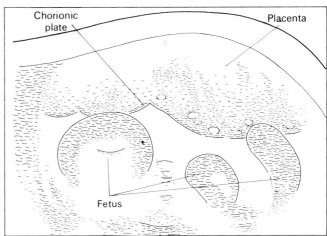

Fig. 12.54 Grade II placenta, longitudinal section. Echogenic densities appear in the basal layer and the texture of the placenta changes with the number of randomly dispersed echogenic densities increasing, and becoming confluent. The undulations in the fetal plate are more obvious with indentations and small 'comma-like' densities. Small cystic areas are also seen in this placenta. They are thought to be of no clinical significance. (Scale 2:5).

 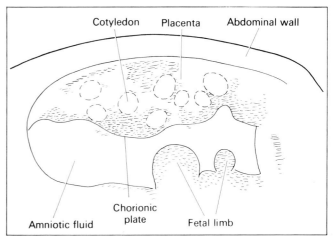

Fig. 12.55 Grade III placenta longitudinal section. This grade of placenta is normally seen after 36 weeks gestation. The indentations in the fetal plate increase, and may communicate with the basal layer. The placental substance changes its characteristics with areas of low-level density (fall-out areas) between linear echogenic densities, which produce a compartmental appearance considered to represent the fetal cotyledons. (Grannum et al 1979, Winsberg 1973, Crawford 1962, Fisher et al 1976).

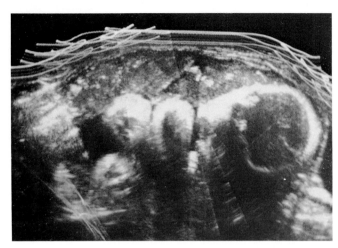 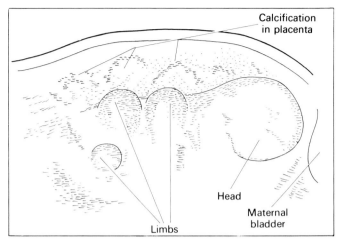

Fig. 12.56 Longitudinal section. Grade II placenta with areas of calcification. Muliple high-density areas are seen in this placenta which represent calcific foci in the intercotyledonary septa. These are associated with degenerative changes. They are rarely seen before the 35th gestational week. (Scale 2:5)

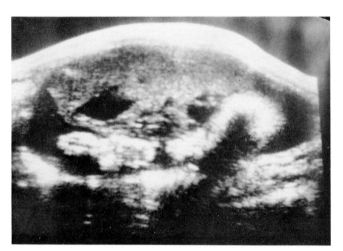 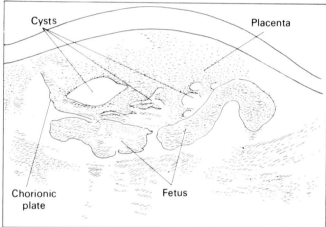

Fig. 12.57 Longitudinal section of placenta with cystic degeneration and subchorionic fibrin deposits. Fibrin cysts are seen at all stages of pregnancy. (Scale 2:5).

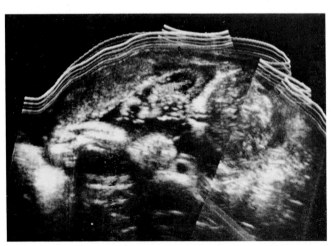 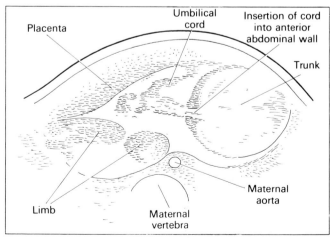

Fig. 12.58 Transverse section. The fetal trunk is sectioned at the level of the insertion of the cord into the anterior abdominal wall. The cord is seen in the amniotic fluid. (Scale 2:5)

OBSTETRIC ULTRASOUND 225

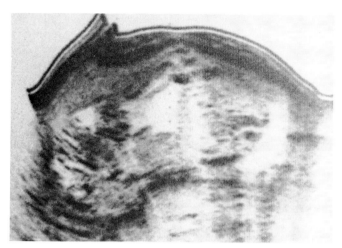 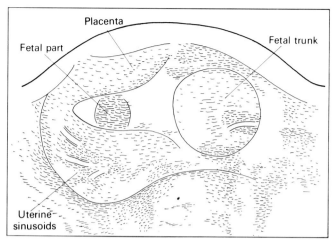

Fig. 12.59 Transverse section. Prominent venous sinusoids are seen in the myometrium deep to the placental tissue.

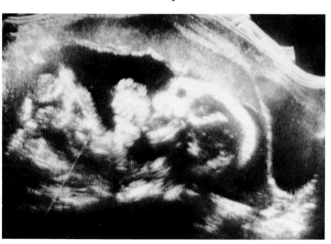 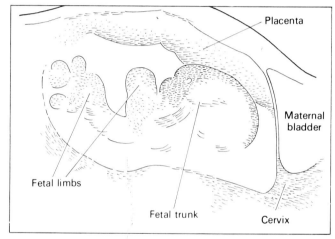

Fig. 12.60 Sagittal section. The placenta is anterior. Its inferior edge is clearly defined above the lower uterine segment.

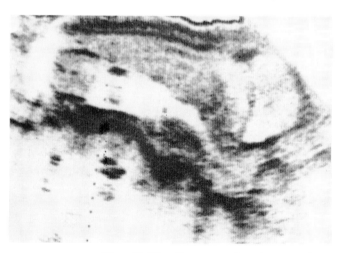 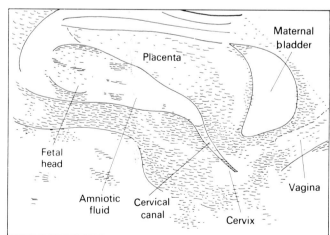

Fig. 12.61 Low anterior placenta. Sagittal section. The placenta extends to the internal os. This may sometimes be produced by a full bladder altering the position of the uterine lower segment or to myometrial contraction.
The diagnosis of placenta praevia should not be made when the bladder is overdistended. Apparent migration had been suggested as a cause for change of placental position as pregnancy progresses. This however is now discounted, and the change in position is considered to be due to elongation of the lower uterine segment as the pregnancy progresses.

226　ULTRASONIC SECTIONAL ANATOMY

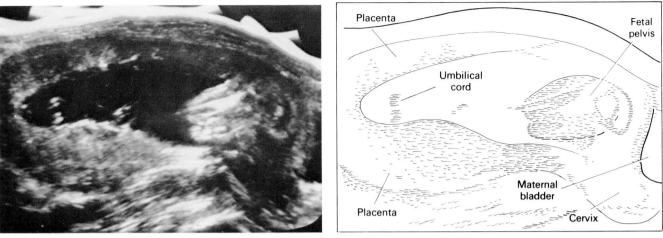

Fig. 12.62　Fundal placenta. Sagittal section.
　　　The placenta is fundal, extending onto both the anterior and posterior wall of the uterus.

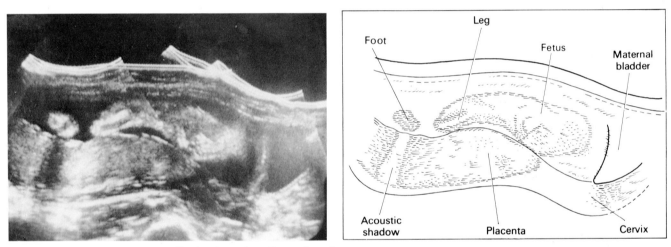

Fig. 12.63　Sagittal section. Posterior placenta.
　　　The placenta is posterior and lateral and relatively high in position.

REFERENCES

American Institute of Ultrasound in Medicine 1976 Standard presentation and labelling of ultrasound images. Journal of Clinical Ultrasound 4: 393

Bom N, Lancee C T, van Zwieten G, Kloster F E, Roelandt J 1973 Multiscan echocardiography. 1. Technical description. Circulation 48: 1066

Campbell S, Wladimiroff J W, Dewhurst C J 1973 The antenatal measurement of foetal urine production. Journal of Obstetrics and Gynaecology of the British Commonwealth 80: 680.

Coleman D J, Konig W F, Katz M S 1969 A hand-operated ultrasonic scan system for ophthalmic evaluation. American Journal of Ophthalmology 68: 256

Coleman D J, Abramson D K, Jack R L, Franzen L A 1974 Ultrasonic diagnosis of tumours of the choroid. Archives of Ophthalmology 91: 344

Coleman D J, Lizzi F L, Jack R L 1977 Ultrasonography of the eye. Lea and Febiger, Philadelphia

Crawford J M 1962 Vascular anatomy of the human placenta. American Journal of Obstetrics and Gynaecology 84: 1543

Donald I 1963 The use of ultrasonics in the diagnosis of abdominal swellings. British Medical Journal 2: 1154

Fisher G C, Garrett D, Kossoff M E 1976 Placental aging monitored by grey scale ultrasonography. American Journal of Obstetrics and Gynaecology 124: 483

Garrett W J 1980 Foetal organ imaging. In: Sabbagha R E (ed) Diagnostic ultrasound applied to obstetrics and gynaecology. Harper Row, New York

Garrett W J, Robinson D E 1970 Foetal heart size measured in vivo by ultrasound. Paediatrics 46: 25

Garrett W J, Kossoff G, Jones R F C 1975 Ultrasonic cross-sectional visualisation of hydrocephalus in infants. Neuroradiology 8: 73

Garrett W J, Kossoff G 1977 Grey scale examination of the brain in children. In: White D, Brown R E (eds) Ultrasound in medicine. Plenum Press, New York, vol 3A, p 821

de Graaf C S, Taylor K J W, Simonds B D, Rosenfield A J 1978 Gray scale echography of the pancreas. Re-evaluation of normal size. Radiology 129: 157

Grannum P A T, Hobbins J C 1979 The ultrasonic changes in the maturing placenta and their relation to foetal pulmonic maturity. American Journal of Obstetrics and Gynaecology 133: 8 915

Kossoff G, Garrett W J, Radovanovich G 1974 Ultrasonic atlas of brain of normal infant. Ultrasound in Medicine and Biology 1: 259

Kossoff G, Carpenter D A, Radovanovich G, Robinson D E, Garrett W J 1975 Octoson, a new rapid multitransducer general purpose water-coupling echoscope. Excerpta Medica Congress Series no. 363: 90

LeMay M 1978 B scan ultrasonography of the anterior segment of the eye. British Journal of Ophthalmology 62: 651

Leopold G R, Asher W M 1971 Deleterious effects of gastro-intestinal contrast media on abdominal echography. Radiology 98: 637

Lutz H T, Petzoldt R 1976 Ultrasonic patterns of space-occupying lesions of the stomach and intestine. Ultrasound in Medicine and Biology 2: 129

McLeod D, Restori M, Wright J E 1977 Rapid B scanning of the vitreous. British Journal of Ophthalmology 61: 437

Mould R F 1972 An investigation of variations in normal liver shape. British Journal of Radiology 45: 586

Mundt H G, Hughes W F 1956 Ultrasonics in ocular diagnosis. American Journal of Ophthalmology 41: 488

Meyers Morton A 1976 In: Dynamic radiology of the abdomen. Springer–Verlag, Berlin

Oksala A 1962 Observations on the depth resolution in ultrasonic examination of the eye. Acta ophthalmologica (KbH) 40: 375

Pape K, Blackwell R J, Cusick G et al 1979 Ultrasound detection of brain damage in preterm infants. Lancet 1: 1261

Piirionen O, Kaihola H L 1975 Uterine size measured by ultrasound during the menstrual cycle. Acta obstetrica et gynecologica scandinavica 54: 247

Popp R L, Fowels R, Coltart D J, Martin R P 1979 Cardiac anatomy viewed systematically with two-dimensional echocardiography. Chest 75: 579

Restori M, Wright J E 1977 C-scan ultrasonography in orbital diagnosis. British Journal of Ophthalmology 61: 735

Rubin C, Kurtz A B, Goldberg B B 1978 Water enema: a new ultrasound technique in defining pelvic anatomy. Journal of Clinical Ultrasound 6: 28

Sample W F 1980 Gray scale ultrasonography of the normal female pelvis. In: Sanders R, James A E (eds) Ultrasonography in obstetrics and gynaecology. Appleton Century and Crofts, New York

Sample W F, Gottesman J E, Skinner D G, Erlich R M 1978 Gray scale ultrasound of the scrotum. Radiology 127: 225

Sample W F, Sarti D A 1978 Computed body tomography and gray scale ultrasonography. Anatomic correlations and pitfalls in the upper abdomen. Gastrointestinal Radiology 3: 243

Sample W F, Sarti D A 1978 Computed tomography and gray scale ultrasonography of the adrenal gland: a comparative study. Radiology 128: 377

Sarti D A, Lazere A 1978 Re-examination of the deleterious effects of gastro-intestinal contrast material on abdominal echography. Radiology 126: 231

Spangen L 1976 Ultrasound as a diagnostic aid in ventral abdominal hernia. Journal of Clinical Ultrasound 3: 211

Tajik A J, Seward J B, Hogler D J, Mair D D, Lie J T 1978 Two-dimensional real-time ultrasonic imaging of the heart and great vessels. Technique, image orientation, structure, identification and validation. Mayo Clinic Proceedings 271

Watanabe H, Igari D, Tanahashi Y, Harada K, Saitoh M 1974 Development and application of new equipment for transrectal ultrasonotomography. Journal of Clinical Ultrasound 2: 91

Weill F, Schraus A, Eisenscher A, Bourgoin A 1977 Ultrasonography of the normal pancreas. Success rate and criteria for normality. Radiology 123: 417

Winsberg F 1973 Echogenic changes with placental aging. Journal of Clinical Ultrasound 1: 52

INDEX

Note: Page numbers printed in **bold** refer to illustrations and captions.

Abdomen
   anterior abdominal wall, 44–**46**, **60**, **81**,
      82, **120**, **123**, **135**, **139**, **145**, **152**,
      157, **158**, **167**, **171**, **177**, **186**, **203**,
      205, **206**, **208**, **210**, **211**, **223**
      muscles, 44, **96**
        see also Rectus abdominis
      sections, **44–46**
   posterior abdominal wall, 52–**57**, **171**
      anatomy, 52
      muscles, 52, **53–55**
        see also Iliacus; Psoas major; Quadratus lumborum
      prone sections, **56–57**
      supine sections, **53–55**
   viscera
      lower abdominal, 117–143
      upper abdominal, 63–98
Acetabulum, **187**
   acetabular region, **60**, **61**
Acoustic artefact *see* Artefacts
Acoustic enhancement, **81**, **86**, **128**, **219**
Acoustic shadow, **81**, **87**, **145**, **150**, **153**, **189**, **204**, **212**, **213**, **220**, **226**
   air, **25**
   calculus, **86**
   catheter, **122**
   duodenum, **83**, **153**, **156**
   gas, **59**, **86**, **89**, **92**, **102**, **146–148**, **151**, **159**, **161**, **162**, **187**
   Lippes loop, **139**
   portal structures, **78**
   ribs, **54**, **57**, **188**, **213**, **216**, **217**
   Rodney Smith tube, **79**
   scar, **46**
   small intestine, **156**
   transverse processes, **56**
   vertebral arch, **56**
   vertebral bodies, **220**
Acoustic window, 8, **95**, 97, **186**
Adnexal mass in pelvis, **111**
Adrenal gland, 97–98
   anatomical relations, 97
   echogenicity, 97
   fetal, **219**
   left, **98**
   right, **98**, **164**
   ultrasonic examination, 97
Allantois, 200
Amnion, 200, **209**
Amniotic fluid, 200, 202, **205**, **209–211**, 220, **223**, **225**
Amniotic membrane, **211**
Ampulla
   of the vasa, 118, 119, **121**
   of Vater, 80, **95**
Aorta
   abdominal, 44, 48–53, 55, 65–67, 72, 74, 75, 83, 88, 89, 91–95, 98, 100, 101, 145–147, 150, 153, 155–157, 165, 166, 168, 170, 171, 174, 175, 177–187, 189, 190, 192, 218, 224
   atheromatous, **178**, **179**
   bifurcation, **121**, **178**, **179**
   branches, 174–175, **177–187**
   course, 174
   dilated, **179**
   fetal, **216**, **218**, **219**
   foramen, 47
   root, 37, 40, 42
   thoracic, 40, 42, 48, 50, 177
      fetal, **216**
   walls, 37, 42
Aortic valve, **37**, **40**, **41**
   cusps, **37**, **40–42**
Aqueduct of Sylvius, 8
Arcuate artery, 100, **107**, **115**, **116**
Arcuate ligament, **180**
Artefacts
   acoustic, **21**, **62**, **79**, **139**
      see also Acoustic shadow
   reverberation, *see* Reverberation
   ringdown phenomena, **62**
Ascites, **45**, **167**
Atrioventricular valves, **41**
Atrium
   left, **37**, **40–42**, **48**, **50**, **51**, **70**, **177**
      appendage, **40**
   right, **40**, **41**, **48**, **50**, **51**, **70**, **187**

Balloon, water-filled, in prostate examination, **121**
'Baum's bumps', **21**
Bile, ultrasonic appearance, 80
Bile duct, common, 80, **87–90**, **94**, **95**, **145**, **169**, **184**, **188**, **191**
 canaliculi, 80
 course, 80
 variants, 80
Biliary tree, ultrasonic appearance, 80
Bladder, urinary, **46**, **60–62**, **110**, **111**, **114**, 117, **119–121**, **123–125**, **133–143**, **145**, **157**, **158**, **160–162**, **171–173**, **186**, **187**
 anatomy, 117
 echogenicity, 117
 fetal, **219**, **220**
 in obstetric examination, 198
  problems, 198
 in pelvic examination, 58
 in transplanted kidney examination, 111
 maternal, in obstetric examination, **202–212**, **215**, **219–221**, **224–226**
 relations, 117
 site in adult and children, 117
 trabeculation, **119**
 ultrasonic examination, xiv, 117
 wall, **173**
Blastocyst, 199, 201, **203**
 stage, **202**
BPD (biparietal diameter) measurement, 198, 202, **206**, 214
 see also Internal biparietal diameter
Brain, 1–11
 B-mode examination, 1
 coronal sections, **6–7**
 cross-sectional echography, 1–7
 image variation with age, 1
 interhemispheric fissure, **9–11**
 lateral sulcus, 9, 10
 transverse sections, **2–3**
  tilted 20° from horizontal plane, **4–5**
Breast, **27–34**
 anatomy, 27
 ducts and glandular tissue, **27–34**
 fibrous tissue, **32**
 in puerperium, **33**
 nipple, 27, **28–34**
 normal, ultrasonic appearance, **28**, **29**
 perimenopausal, **31**
 postmenopausal, **32**
 post partum, **34**
 premenstrual, **30**
 suspensory ligaments, **28**, **29**
 transverse and longitudinal sections, **28–34**
Breech presentation, **221**
Broad ligament, 131, 132
B-scan (static images), xiii
 brain, 1

cerebral ventricles, 8–9, **9–10**
diaphragm, **50**

Calculi, in gall bladder, 80, **86**
 acoustic shadow, **86**
Carotid artery
 common, **23–26**, **195–197**
  bifurcation, **197**
  course, 193
  left and right, course, 193
 external, **197**
  course, 193
 internal, **196**, **197**
  course, 193
Carotid sheath, 23
Carotid sinus, 193, **197**
Catheter, urinary, ultrasonic appearance, **122**, **125**
Cavum septi pellucidi, **9**, **11**
 fetal, **214**
 see also Septum pellucidum
Cerebellum, 1, **11**
Cerebral aqueduct, **10**, **11**
Cerebral midline, in fetus, 202, **213–215**
Cerebral peduncles, fetal, **215**
Cerebral ventricles, 1, **8–11**
 anatomy, 8
 coronal sections, **9–11**
 fourth ventricle, 8
 horns, **8–11**
 in fetus, 202, **213–215**
 lateral, 8
 lateral ventricular ratio, 8, **213**
 midline sagittal section, **10**, **11**
 parasagittal section, **10**, **11**
 third ventricle, 8, **10**, **11**
Cerebrospinal fluid, 8
Cervical canal, **133–135**, **203**, **225**
Cervix, **46**, **123**, **133**, **135–137**, **139**, **171–173**, **203**, **205**, **206**, **209**, **225**, **226**
Chest wall, **83**, **95**, **149**, **168**, **170**
Chordae tendineae, **42**, **43**, **50**
Chorion frondosum, 201
Chorionic (fetal) plate, **207**, **210**, **222–224**
Chorionic villi, 199, 201, **203**, **204**
Chorion laeve, 200, 201, **203**
Choroid
 of eye, **15–17**, **19**
 plexus, 8, **11**
Ciliary body (pars corona ciliaris), **15**
Cingulate gyrus, **9–11**
Coeliac artery, **178**
Coeliac axis (trunk), **75**, **88**, **92**, **93**, **146**, **166**, **168**, **177**, **180**
 course, 174
Colon, **47**, **69**, **76**, **81**, **82**, **101**, **107**, **147**, **148**, **151**, **159**, **166**, **170**, **188**
 ascending and descending, peritoneal reflections, **164**

Colon *(contd)*
   hepatic flexure, **159, 165**
   pelvic, **58**
   sigmoid, **160**
   splenic flexure, **147, 164**
Condyles of femur and tibia, **195**
Connecting stalk, 200
Contact echoscope, 1
   B scan, 9, **9**, **10**
   real time, 23
Copper 7 in uterus, ultrasonic appearance, **139**
Cornea, **14–16, 18–21**
   ultrasonic examination, 12
Coronary artery
   left, main, **40, 41**
   right, **37, 40**
Coronary ligament, **164, 166**
Corpus callosum, **9–11**
Corpus cavernosum, **128, 129**
Corpus luteum, 132, **142**
   cyst, **143, 208**
Corpus spongiosum, **128, 129**
Cricoid cartilage, **25**
CRL (crown rump length) measurement, 202, **204–207**
Crura *see* Diaphragm
'C' scan of orbit, **22**
Cumulus oöphorus, **142**
Cyst
   corpus luteum, **143, 208**
   in placenta, **224**
   lower abdominal, **185**
   pouch of Douglas, **208**
   retention, **203**
Cystic artery, 80
Cystic duct, 79, **85**

Dartos muscle, scrotum, **127**
Decidua, **135**, 199–200, **202**
   basalis, 199, **205, 206**
   capsularis, 199, 200, 202, **205, 206**
   parietalis, 199, 200, **203**
Depth markers, explanation, xvi
Diaphragm, **46–53, 65–69**, 78, 79, 82, 90, 92, 98, 107, 108, 110, 146, **148–151, 154, 159, 166, 167, 170, 177, 180, 183, 187, 188, 191**
   anatomy, 46–47
   aortic hiatus, 47, **51**
   central tendon, 46–**50**
   crura, 44, **46–49**, 51, 76, 93, 96, 98, 181, 182, 184, 187, 188, 192
   double insertion, **69**
   everted, **50**
   foramina, 47
   left and right domes (hemidiaphragms), 46, **47, 49, 65, 69, 70, 77, 86, 102, 148–150, 167, 170, 188**
   movement (excursion), 47, **52**
   sections, **47–52**
Doppler, xiv–xv
   lower limb vessel examination, 192, **194, 195**
   neck vessel examination, 192, **197**
   showing carotid artery stenosis, xv
Ducts *see* Bile duct; Breast, ducts; Hepatic duct; Pancreas, duct
Ductus
   deferens, 117, 125
   venosus, **218**
Duodenum, **76**, 84, 88, **89**, 93, **94, 101, 153–157, 164, 171, 182, 184, 185**
   acoustic shadow, **83, 89**

Echo format, white and black, 64
Echocardiography
   cross-sectional, 35–36
   ideal subject, 35
   M-mode, 35
   optimum probe position, 35
   sector scanning, 35–36
Embryo
   form at different ages, 201–202
   formation, 200
Embryology, 198–201
Emphysema, appearance of diaphragm in, **50**
Endocervical canal (cervical canal), **133–135**, 203, **225**
Endometrium, uterine, **131**
   echogenicity in gestation, **202**
   endometrial cavity, **134, 136**
   thickened, **135, 136**
Epididymis, 125, **127, 130, 131**
   echogenicity, 126
Epigastrium, **44, 45, 50, 189**
Erector spini, **55, 145**
Expiration *see* Diaphragm, movement
Extraembryonic coelom, 199, 200, **203**
Extraperitoneal space, right, 163
Eye, 12–22
   anterior segment (chamber), **14–16, 18–20**
   angle, 15
   artefactual shortening, **21**
   axial length
      abnormal, **20**
      normal, **13**
   examination, 12–13
   horizontal section, **16**
   iris, **14–16, 20**
   lens, **14–16, 18–21**
   posterior wall, **16, 17, 21**
   sclera, **15–17, 19**
   see also Cornea; Orbit; Retina; Sclera
Eyelids, **14, 16, 18–20**
   interference in ophthalmological examination, 12–13

Falciform ligament, **45**, **63**, **167**
Fallopian tube, **136**, 198, 199
Fascia
  anterior abdominal wall, 44
  pelvic, 58
  thoracolumbar, **55**
Fat
  extraperitoneal, **44**, **45**
    in pelvis, 58, **62**
  in abdominal fascia, 44
  in anterior abdominal wall, **44–46**
  in breast, 27–**32**
  in buttock, **56**
  orbital, **14**, **17–22**
    herniated, **20**
  perinephric, **96**, 99, **101**, **105**, **106**, **108**
Femoral artery
  common, **179**, **194**, **195**
  course, 192
  deep, **179**, **194**, **195**
    course, 192–193
  lateral circumflex, **195**
  superficial, **179**, **194**, **195**
Femoral head *see* Femur, head
Femoral nerve sheath, **60–62**
Femoral vein, **179**, **194**
  course, 193
Femur, **195**
  head, 55, **61**, **179**, **187**
Fetal plate (chorionic plate), **207**, **210**, **222–224**
Fetus, **204–226**
  bowel, **217**
  foot, **207**, **222**, **226**
  head and face, **205–207**, **209–215**, **217**, **221–225**
  heart, **215**, **216**, **219**, **221**
    movements, 202
  limbs, **205**, **206**, **211–213**, **215–220**, **223–226**
  lung, **215**, **216**
  membrane, 200
  neck, **221**
  pelvis, **226**
  ribs, **212**, **213**, **216**
  shoulder, **212**
  spinal canal, **206**, **211–213**, **220**
  spine, 202, **207**, **211–213**, **215**, **216**, **218–222**
  trunk, **206**, **210–213**, **215–217**, **220**, **221**, **224**, **225**
  twin, **210–211**
  ultrasonic anatomy, **211–221**
  use of term 'fetus', 202
Fibroid, uterine *see* Fibromyoma
Fibromyoma, uterine, **186**, **208**, **209**
First trimester of gestation, ultrasonic image, 201–202, **202–209**
Follicle, ovarian, **140**, **142**
  development during menstrual cycle, **142**
  mature, **123**

Fontanelle, as acoustic window, 8
Foramen
  of Monro, 8, **10**, **11**
  magnum, fetal, **212**
  of Winslow, 80, **88**, 163, **164**, **168**, **169**
'Four-chamber' view of heart, 35, 36, **41**, **50**
Fourth ventricle, 8
Frequency used in ultrasonic examination, xiii, xiv
  bladder examination, 117
  breast, 27
  eye, 12
  liver, 64
  thyroid and neck, 23
  transplanted kidney, 111

Gall bladder, **54**, **68**, **72**, 79–**89**, 93, 98, **101**, **103**, **106**, **107**, **109**, 153, **155**, 159, **165**, **171**, **181**, **182**, **184**, 192
  anatomy, 79
  anomalies, 79
  calculus in, **86**
  fetal, **219**
  fluid level, **85**
  infundibulum, **85**
  junctional fold, **84**, **85**
  kinked, **84**
  neck, **85**
  'Phrygian cap' deformity, **84**
  septum, **84**, **155**
  shape and size, 79, 80
  site, 79–80
  ultrasonic examination, 80
  wall, **85**
Gastroduodenal artery, 80, **87**, **88**, **90**, **94**, **169**, **181**, **182**
Gastro-intestinal tract, 144–**162**
  appearance, 144, **145**
  contents, 144
  gaseous distension, 145
  technique of examination, 144
Gastrosplenic ligament, **164**
Germ disc, 199
Gestation
  sac, 201–**204**, **207**, **208**, **210**, **221**
  ultrasonic appearance and embryonic form at different ages, 201–202
    4 weeks, **202**
    4+ weeks, **203**
    6 weeks, **203**
    6–7 weeks, **204**
    7 weeks, **208**
    8 weeks, **207**, **208**, **210**
    9 weeks, **221**
    9+ weeks, **205**
    10 weeks, **205**
    11 weeks, **206**
    12 weeks, **211**
    13 weeks, **206**, **210**

Gestation *(contd)*
    14 weeks, **206–208, 212, 222**
    16 weeks, **209, 212, 213**
    18 weeks, **213**
    22 weeks, **210**
    34 weeks, **220, 221**
Glisson's capsule, 64, **66, 69, 74, 77, 86**
Gluteal muscles, **56, 60, 161**
Great vessels, long-axis views, 36
    see also Aorta; Pulmonary artery, vein; Vena cava
Grey scale ultrasound, xiii

Hartmann's pouch, 79
Heart, 35–**43, 47, 49, 50, 65–67, 70, 76, 82, 146, 150, 167, 177**
    cross-sectional examination, 35–36
    diastole, **37, 38, 40, 42**
    disordered spatial arrangements, 35
    end-systole, **37, 39, 41**
    four-chamber view, 35, 36, **41, 50**
    long-axis scans, **42**
    mid-systole, **42, 43**
    motion, 35
    parasternal probe position, **37–40**
    short-axis scans, **38–40, 43**
    subxiphoid probe position, **41**
Hemidiaphragm *see* Diaphragm, left and right domes
Hepatic artery, 80, **86–88, 91, 93, 101, 166, 168, 169, 180, 181, 183, 191, 192**
    branches, 63
    common, 63
    variants, **76,** 80
Hepatic duct, 80, **86–88, 166, 168**
Hepatic vein, **47, 48, 51, 54,** 64, **67, 68, 70, 71, 73, 74, 78, 81, 82, 85–87, 90, 91, 98, 151, 166, 169, 180, 181, 183, 187, 191**
    course, 175
    branches, 64, **68, 73, 74, 76, 77, 188, 189**
Hepatoduodenal ligament, **86**
HFUPR (hourly fetal urine production rate), **219**
History of ultrasound, xiii–xv
Horns of cerebral ventricles
    anterior, **9, 10**
        fetal, **213, 214**
    inferior and posterior, 8, **11**

Iliac artery
    common, **178, 179, 185, 186**
        course, 174–175
    external, **186, 187, 194**
        course, 175
    internal, **141, 187**
        course, 175

Iliac fossa, **55, 59, 112**
Iliac spine, **60**
Iliacus, **55, 59, 60, 112, 113, 123, 124, 186**
Iliac vein, external, **187, 194**
Iliac vessels, **61, 173, 186**
Iliopsoas, **46,** 55, **60–62, 123, 140, 172, 173, 187, 194, 208**
Ilium, **55, 56, 59–61, 112, 123, 124, 172, 173**
Imaging, bistable, xiii
Individual differences, anatomical, xv
Inferior mesenteric artery, **179**
    course, 174
Inferior mesenteric vein, course, 176
Inferior oblique (eye), **17, 19**
Infracolic space, **164**
Inframesocolic space, **171**
Inspiration *see* Diaphragm, movement
Interatrial septum, **40, 41, 50, 51**
Internal biparietal diameter, 8
    and LVR, 8
Interventricular foramen (brain), 8
Interventricular septum (heart), **37–39, 41–43, 50, 70, 177**
    fetal, **216**
Intervillous spaces, 201
Intraperitoneal space
    left anterior, 163, **167, 168**
    left posterior, 163, **169–171**
    right anterior, 163, **164, 166, 167**
    right posterior, 163, **165, 166, 169**
IUCD, ultrasonic appearance, **139, 173, 207**

Jejunum, **159**
Jugular vein, internal **24–26, 195–197**
    course, 193
    relationship to common carotid artery, **25**

Kidney
    anatomy, 99
    anomalous, 99, **107–111**
    calyectasis, **114**
    calyx, **106, 107, 110, 115**
    capsule, 99, **107**
    cortex, 99, **106, 107, 115, 116**
    cortical columns, **107, 115, 116**
    crossed ectopic, **108**
    dimensions
        anatomical, 99
        radiographic, 99
    duplex, **110, 111**
    echogenicity, 100
    ectopic presacral, **110**
    fetal, **218, 219**
    horseshoe, **109**
    infundibulum, **106**
    left, **49, 51, 52, 54, 57, 65,** 92, **93, 96–98, 100–105, 109, 148–150, 167, 170, 181, 183, 189**

Kidney *(contd)*
    malrotated, **108, 110**
    medulla, 99, **106, 116**
    normal variants, 100
    normotopic, 99–**107**
    pancake, **110**
    pelvic, **110, 111**
    right, **47, 56, 57, 68, 69, 72, 73, 77–79, 81–86, 93, 94, 98, 100–109, 153–155, 164–166, 181–185, 188, 190, 192**
    transplanted, 111–**116**
        configuration, 111
        operative technique, 111
        precautions for ultrasonic examination, 111
        rotation, 111
        site, 111
    ultrasonic examination, 99–100
    see also Renal

Larynx, 25
Lateral cerebral sulcus, 1, **9, 10**
Lateral cerebral ventricles, 8
    anterior horn, 8, **9, 11**
    posterior horn, **11**
    temporal horn, **11**
Lateral orbital wall, **14, 17, 18, 20**
Lens of eye, **14–16, 18–21**
    acoustic artefact from, **21**
Lesser omentum, 63
Lesser sac (left posterior intraperitoneal space), **49, 163, 164, 169–171**
Ligament
    falciform, **45, 63, 167**
    right lateral, **166**
    round, **136**
Ligaments of Cooper, **28, 29**
Ligamentum teres, 63, **93, 95, 100, 153, 165, 168, 170, 182, 183**
    fissure, **72**
Ligamentum venosum, 63
    fissure, **66, 151**
Limb presentation, **220**
Limited sector scan, ovary, **142**
Linea alba, **44, 46**
    anatomy, 44
Linear array, 9
    scans, **35, 188, 214, 215**
    'sequenced', xiii–xiv
Lippes loop in uterus, ultrasonic appearance, **139**
Liquor amnii, 200, 202, **205, 209–211, 220, 223, 225**
Liver, **44, 45, 47–54, 63–79, 81–95, 98, 101–103, 106–110, 145–157, 159, 165, 167–171, 177–184, 187–192**
    amount examinable by ultrasound, 64
    anatomy, 63
    as transonic window, 97
    bare area, 163
    caudate lobe, **48, 66, 67, 71, 72, 151, 152, 180, 190, 191**
    fetal, **218, 219**
    left lobe, **49, 51, 65–67, 71, 72, 74, 75, 79, 92–94, 148, 149, 166–168, 170, 180, 183**
    lobar fissure, main, **81, 85**
    quadrate lobe, **48, 66, 67, 71, 72, 151, 180**
    right lobe, **47, 51, 68–75, 83, 93, 100–103, 107, 165–167, 180**
    sagittal sections, **65–70**
    site, 63
    size, 64
    transverse sections, **70–73**
    ultrasonic examination, 64
    variant lobes, 64
    vessels, 63–64, **69, 70**
    see also Hepatic artery, duct, vein; Portal vein
Lobes *see* Liver; Prostate; Thyroid
Lower limb vessels, 192–193, **194–195**
Lumbar spine, **59**
    transverse processes, **54–56, 59**
    vertebrae, **56–58**
Lung, **47**
    fetal, **215, 216**
LVR (lateral ventricular ratio), 8, **213**

Mammary gland *see* Breast
Medial orbital wall, **19**
    deficient, **20**
Median arcuate ligament, 46–47, **48, 51**
Mesentery, root of, 163, **164**
Mitral valve, **38, 41, 42, 50**
    leaflets, **37, 38, 41–43**
    orifice, **38**
Morison's pouch, 163, **164, 169**
Morula, 199
Myelinated nerve tracts, 1
Myometrium, uterine, 131, **133, 136, 224**
    contraction of, **209**
    thickened, **135**
Myopia, axial, **20**

Neck
    acoustic density, 23
    fetal, **221**
    muscles, 23–**26, 195, 196**
    neurovascular bundle, **24, 25**
    soft tissues, 23–**26**
    vessels, **24–26,** 193, **195–197**
Nuclei, caudate and thalamic, **9–11**

Obstetric examination by ultrasound, technique, 198
Obturator
  foramen, **61**
  internus, **61, 137, 172**
Oesophago-gastric junction, **146, 177**
Oesophagus, 47, **49, 177**
Open water tank technique, 126, **127**
  see also Water bath
Ophthalmological ultrasound instruments, xiv
  focussing, 12
  probes, 12
Optic nerve, 1, **14, 17–19, 21, 22**
  dimensions, 17
  movements, 18
    with abduction, **18**
    with adduction, **19**
Orbit, 12–22
  'C' scan (coronal scan), **22**
  examination, 12–13
  fat, **14, 17, 20, 22**
  lateral wall, **14, 17, 18, 20**
  medial wall defect, **20**
  transverse scan, **14**
Ovary, 132, **140–143**
  change during menstrual cycle, 132
  cumulus oöphorus, **142**
  ligaments, 132
  relations, 132
  retention cyst, **136, 143, 208**
  size, 132
  stromal texture, **140**
  ultrasonic examination and appearance, **123**, 132, **136, 140–143, 161, 172, 187, 208**
Ovum, 198
  fertilised, development, 199
    implantation, 199

Pancreas, **57, 67, 72, 79, 83, 88–95, 101,** 152, 153, **155, 165, 171, 181, 182, 184, 191**
  anatomy, 89
  ducts, 89–90, **95**
  echogenicity, 90
  head, **87, 90, 91, 93, 169, 192**
  landmarks, 90
  neck, **91, 94, 169**
  shape, 89
  size, 90
  tail, **51,** 89, **92–94, 101, 104**
  ultrasonic examination, 90
Papillary muscles (heart), **37, 38–39, 41–43**
Paracolic gutters, 163, **164, 171**
Paracolic space, left, **164**
Parathyroid glands, 24
Paravertebral muscles, **108**
Paravesical fossae, 164
Pars corona ciliaris (ciliary body), **15**

Pars plana, **15**
Pelvis
  anatomy, 57–58
  connective tissue, **119, 120**
  fascia, 58
  floor, **119, 137, 161**
  muscles, **61, 62, 123, 140, 172**
    anatomy, 58
  peritoneum, 58, 163–164
  pseudomasses, 144
  sections
    greater pelvis, **59–62**
    lesser pelvis, **58, 60–62, 186**
  skeletal boundaries, **57–61**
  viscera, 58, **117–143**
Penis, **126–128**
  dorsal vein, **128**
  fascia, **128**
  root, **128**
  tunica albuginea, **128**
  ultrasonic examination
    using conventional contact scanner, **130–131**
    using UI Octoson, **126–129**
Pericardial effusion, **82**
Peritoneum
  compartments (recesses), **162–173**
    anatomy, 162–163
    boundaries, 163–164
  covering liver, 63
  parietal, 44
  pelvic, 58, 163–164
  reflections, **164**
  see also Intraperitoneal space
Phased array systems, xiv
  echocardiography, 35
Phrenicocolic ligament, **164**
Piriformis, **61, 62, 123, 140, 172**
Placenta, 202, **205–213, 215, 217–226**
  calcification, **224**
  cotyledons (lobes), 201, **223**
  cystic degeneration, **224**
  development, 201
  fundal, **226**
  Grade 0, **222**
  Grade I, **222**
  Grade II, **224**
  Grade III, **223**
  placental septa, 201
  posterior, **226**
  praevia, 198, **225**
Planes of section, xvi
  transverse orbital, in ophthalmological examination, 13
Pleural effusion, **102**
Pons, **11**
Popliteal artery, **195**
  course, 193
Porta hepatis, 51, 63, **71, 72, 74, 79, 95, 100, 145, 168, 170**
Portal system, extrahepatic, 176, **190–192**

Portal triad, 176
Portal vein, **45, 51, 53, 54, 63, 64, 67–69, 72, 75–78, 81, 82, 84–91, 93, 98, 101, 110, 145, 147, 151, 153, 156, 166, 169, 180–184, 187–189, 191, 192**
   course, 176
   left and right branches, 64, **65, 66, 68, 69, 74–78, 87, 88, 166, 168, 182, 183, 188, 191, 192**
Pouch of Douglas, 131, **143**, 164, **171, 172**
Prevertebral muscles (neck), **24–26, 195**
Probe positions, echocardiography, 36
Problems in ultrasonic examination
   photographic, xiv
   technical, xiii, xv
Profunda femoris (deep femoral artery), **179, 194, 195**
   course, 192–193
Prostate
   anatomy and relations, 118
   capsule, **122**
   echogenicity, 118
   hypertrophy, **119, 120, 122**
   lobes, 118
   size, 118
   ultrasonic appearance, 118
     hypertrophic, **119, 120, 122**
     normal, **119, 121, 157, 160, 173**
   ultrasonic examination, xiv, 118, **119–122**
Pseudomasses (pseudotumours), 144, **147, 148, 157, 160**
Psoas major, **53–55, 57**, 81, 83, **102, 107, 186**
Pubic bone, **187, 194**
Pulmonary artery, **40, 41**
Pulmonary valve, **40**
Pulmonary vein, **41, 48**

Quadratus lumborum, **55–57, 102, 103**

Radial scanning, 118
   rectal scans, **121, 122**
'Real time' scanning
   development, equipment, xiii–xiv
   in breast examination, 27
   in heart examination, 35
   in ophthalmological examination, 13
   in thyroid examination, 23
   mobile systems, 9
   sector scans, cerebral ventricles, **10, 11**
Recto-uterine space *see* Pouch of Douglas
Rectovesical pouch, 164, **173**
Rectum, 58, **61, 62**, 120–122, **136, 137, 160–162**
   gas in, **162**
Rectus abdominis, **45, 46**, 55, 59, 60, **119, 120, 123, 133, 140, 171–173, 185**
Rectus muscles of eye, lateral and medial, **14, 15, 17–21**
Rectus sheath, **45, 46**
Reidel's lobe, 64, **69, 70**
Reid's base line, 1
Renal
   anomalies, 99, **107–111**
   artery, **76, 83**, 98, **171, 182–184, 189, 192**
     course, 174
   capsule, 99, **107**
   fascia, 99, **101, 105, 106**
   hilus, 99, **106**
   hyperplasia, **107**
   length, 100
   parenchyma, 100, **101, 106, 108, 112–115**
   pelvis, 100, **105, 106, 110, 112–114, 184, 188**
     distension, 100, **113, 114**
   pyramids, **106, 107, 115, 116**
   sinus, 99, **101, 105–108, 112–116**
   veins, **73, 75, 94, 165, 182–184, 188–190**
     course, 175–176
   see also Kidney
Resolution of scans, xiii
   ophthalmological, 12, 13
Retina, **16, 17, 19**
Reverberation, acoustic artefact, **25, 46, 47, 62, 73, 97, 102, 104, 120, 136, 139, 150, 165, 166**
   from gas, **119, 145–148, 157**
Ribs, **54, 57, 188, 213, 216, 217**
Ringdown phenomena, **62**
Round ligament, **136**

Sacral promontory, **58, 112**
Sacrospinalis, **54, 56, 57, 96, 103, 104, 108, 186**
Sacrum, **58–60, 62, 123, 136, 205**
   anterior surface, **55**
   fetal, **212**
Scalene muscles, **24, 26, 195, 196**
Scar tissue, in anterior abdominal wall, **46**
Sclera (eye), **15–17, 19**
Scrotum, **125–131**
   fetal, **220**
   ultrasonic examination, 126
     with conventional contact scanner, **130–131**
     with UI Octoson, **127–129**
Sector scans
   cerebral ventricles, **10, 11**
   echocardiography, 35–36
   fetus, **216**
   in ophthalmological ultrasound, xiv, 13
   phase-steered, 9
Seminal vesicles, 118–119
   anatomy, 118
   ultrasonic appearance, 118, **120, 121, 173**
Septum pellucidum, 8, **10, 32, 214**

'Sequenced linear array', xiii–xiv
'Single pass' technique, xiii
Skull thickness and ultrasonic image, 1
Small intestine, **157–159**
   fluid distension, **158**
   folds, **158**
   gaseous distension, **145, 146**
   in pelvic examination, 58
   mucosal pattern, **159**
   shadowing, **156**
Small parts scanner, xiv, **197**
Sphincter of Oddi, 80
Spleen, 49, **95–98**, 100, 103, 104, 147, 149, 150, 164, 167, 168, 170, 181
   anatomy, 95
   hilum, **95, 168, 170**
   ultrasonic appearance and examination, 95
Splenic artery, **91–93**, 150, 181
Splenic recess, **49**
Splenic vein, 51, 65, 75, **91–94**, 101, 150, 153, 177, 178, **182–184**, 192
   confluence with superior mesenteric vein, **91**
   course, 176
Static image *see* B-scan
'Static' scan, xiii
   abdominal examination, xiv
   breast examination, 27
   cerebral ventricles, 8–9
   chest examination, **50**
   kidney examination, 111
   ophthalmological examination, 13
   thyroid examination, 23
Sternocleidomastoid, **24–26, 195–197**
Stomach, 45, 49, 65, 71, **93–97**, 100, **102–104**, 108, **146–150**, 164, 167, 168, 170, 180, 192
   antrum, 66, 67, **151–154**, 177, 178, **190, 191**
   empty, **150, 151**
   fetal, **215, 217, 218**
   fluid in, **147, 168, 170, 215**
   gaseous distension, **145, 148**
   gastric tube, **147**
Strap muscles, **24, 25, 195**
Subhepatic space, right (right posterior intraperitoneal space), 163, **165, 166, 169**
Subphrenic space
   left (left anterior intraperitoneal space), 163, **167, 168**
   right (right anterior intraperitoneal space), 163, **164, 166, 167**
Subsartorial canal, **195**
Superior mesenteric artery, 48, 51, 75, **88, 89, 91, 93, 94**, 101, **145, 155, 156**, 177, 178, **182–185, 189, 190, 192**
   course, 174
Superior mesenteric vein, 48, 49, 66, 67, 72, 79, 83, 88, 89, 91, 94, 95, 101, 152, **153, 155, 165, 169, 171, 177, 182–185, 190, 191**
   confluence with splenic vein, **91**
   course, 176
Supine hypotension in late pregnancy and ultrasonic examination, 198
Swept gain *see* TGC
Sympathetic trunk, **24**

'Target lesion', 144, **146, 151, 152, 156**
Tenon's capsule, **16**
Testis
   echogenicity, 126
   examination
      with conventional contact scanner, **130–131**
      with UI Octoson, **127–129**
   fascial covering, 125
   median raphe, **130**
   size, 125
Thalamus, **9–11**
   fetal, **214**
Third ventricle, 8, **10, 11**
   fetal, **214, 215**
Thoracolumbar fascia, **55, 56**
Thyroid, **23–26**, 195
   acoustic density, 23
   attachments, 23
   cartilage, **25**
   development, 23
   isthmus, **24**
   lobes, 23, **24, 25**
   sagittal section, **25, 26**
   'texture', **25, 26**
   transverse sections, **24, 25**
   ultrasonic examination, 23
TGC (time gain compensation, swept gain), xiii, 64
Trachea, 23, **24, 25**, 195
Transducer, xiii, xiv
   in anterior abdominal wall examination, 44
   in thyroid examination, 23, **26**
   in transplanted kidney examination, 111
   positions in echocardiography, 35–36
   site
      in bladder examination, 117
      in prostate examination, 118
      ultrasonic appearance, **121–122**
   see also Frequency
Transverse mesocolon, 162
Tricuspid valve, **40, 41, 43, 50**
Trophoblast, 199, 202
Trophoblastic shell, 199, **204, 205, 208, 221**
Trophoderm, 199
Twin pregnancy, 198
   ultrasonic appearance, **210, 211**

UI Octoson, 1
   examination of penis and scrotum, 126–**129**
   open water tank technique, 126
Umbilical cord, 200, **218**, **219**, **224**, **226**
   insertion into fetal abdomen, **224**
Umbilical vein, **218**
Umbilical vesicle, 200
Uncinate process, **83**, **91**, **93**, **169**
Ureter, 117–118, **124**, **141**, **187**
   course and relations, 117
   in female, 117–118
   in male, 118
   ultrasonic examination, 118
Ureteric orifices, **124**, **125**
Urethra, **120**, **122**, **128**, **129**
   course in male, 126
Uterovesical pouch, **158**, **164**, **172**, **173**
Uterus, **58**, **61**, **62**, **123**, **124**, 131–132, 133–**136**, 138–**141**, **143**, **158**, **161**, **162**, **171**–**173**, **186**, **202**–**213**, **215**–**226**
   anatomy, 131
   'cavity', **135**
   decidua, **135**, 199–200, 202
   duplex, **138**
   endometrium, 131, **135**, **136**
   fundus, 131, **133**
   ligaments, 131, 136
   myometrium, 131, **133**, **135**, **136**, **209**, **225**
   puerperal, **135**
   relations and site, 132
   retroverted, **133**, **134**, **158**, **207**
   senile atrophic, **134**
   sinusoids, **225**
   size, 131
      change with age, 132
      change with pregnancy, 132
   ultrasonic examination, 132, **133–143**
   uterine tubes, 131

Vagina, **58**, **123**, **133**–**137**, **139**, **158**, **161**, **162**, **171**–**173**, **203**, **205**, **209**, **225**
Vagus, 24
Valsalva manoeuvre
   to visualise inferior vena cava, 175
   to visualise internal jugular vein, **196**
Valves of Heister, 79
Variation, anatomical, xv
   breast, 27
   gall bladder, 79

kidney, 99, 100, **107–111**
liver, 64
Vena cava, inferior, **45**, **47**, **47**, **48**, **51**, **53**, **67**, **70**–**77**, **82**–**84**, **87**–**91**, **93**, **94**, **98**, **101**, **102**, **110**, **145**–**147**, **151**–**153**, **155**, **156**, **164**–**166**, **168**, **169**, **171**, **180**–**192**
   course, 175
   fetal, **215**, **217**
   normal pulsation, **188**
   ultrasonic appearance, 175
Ventricle (heart)
   left, **37–43**, **49**, **50**, **70**, **177**
   right, **37–43**, **48**, **50**, **51**, **70**
      trabeculation, **41**
Vertebral bodies, **53–55**, **57**, **58**, **98**, **146**, **147**, **149**, **185**, **195**
   fetal, **220**
Vertebral canal, **53**, **54**, **56**, **145**
   fetal, **206**, **211–213**, **220**
Vertebral column (spine), **47**, **48**, **51**, **52**, **57**, **65**, **67**, **70**, **71**, **73**, **75**, **83**, **88**, **95**, **96**, **109**, **155**, **157**, **165**, **166**, **168**, **169**, **171**, **178–181**, **183–185**, **192**, **224**
   fetal, **207**, **211–213**, **215**, **216**, **218–222**
   lumbar transverse processes, **54–56**, **59**
Vertebral disc, **54**
Vertebral vessels (neck), **195**
   course, 193
Vertex presentation, **215**
Videotape recording, 35
Vitello-intestinal duct, 200
Vitreous humour, **14**, **16–21**

Water bath
   breast examination, 27
   gall bladder examination, 80
   liquid used, 12
   neck vessel examination, **195–196**
   ophthalmological ultrasound, xiv, 12
   testis examination, **130–131**
      with polythene membrane, 126, **128–129**
   thyroid and neck examination, 23, **195**
Water-delay echoscope, 1
Water enema, 144, **160**
Water load, oral, **150**

Yolk sac, 200